SCIENTIFIC AND TECHNICAL ASSESSMENTS
OF ENVIRONMENTAL POLLUTANTS

Chloroform, Carbon Tetrachloride, and Other Halomethanes:
An Environmental Assessment

A report prepared by the
Panel on Low Molecular Weight Halogenated
Hydrocarbons of the Coordinating Committee for
Scientific and Technical Assessments of
Environmental Pollutants

Environmental Studies Board
Commission on Natural Resources
National Research Council

NATIONAL ACADEMY OF SCIENCES
Washington, D.C.   1978

NOTICE

The project that is the subject of this report was approved by the Governing Board of the National Research Council whose members are drawn from the Councils of the National Academy of Sciences, the National Academy of Engineering, and the Institute of Medicine. The members of the Committee responsible for the report were chosen for their special competences and with regard for appropriate balance.

This report has been reviewed by a group other than the authors according to procedures approved by a Report Review Committee consisting of members of the National Academy of Sciences, the National Academy of Engineering, and the Institute of Medicine.

This study was supported by
the Office of Health and Ecological Effects, Office of Research and Development, U.S. Environmental Protection Agency, Contract No. 68-01-3253.

Library of Congress Card Catalog Number 78-58464
International Standard Book Number 0-309-02763-2

Available from
Printing and Publishing Office
National Academy of Sciences
2101 Constitution Avenue
Washington, D.C. 20418

Printed in the United States of America

PANEL ON
LOW MOLECULAR WEIGHT HALOGENATED HYDROCARBONS

Julian B. Andelman (Chairman), University of Pittsburgh
John H. Cumberland, University of Maryland
Edward D. Goldberg, University of California, San Diego
David G. Hoel, National Institute of Environmental
   Health Sciences
Robert J. Moolenaar, Dow Chemical Company
Reinhold A. Rasmussen, Oregon Graduate Center

Consultants and advisors:

David Pierotti, Oregon Graduate Center
Mathew J. Schon, D. Chester Engineers

Staff Officer:

Adele L. King, Environmental Studies Board

COORDINATING COMMITTEE FOR
SCIENTIFIC AND TECHNICAL ASSESSMENTS
OF ENVIRONMENTAL POLLUTANTS

Ian C.T. Nisbet (Chairman), Massachusetts Audubon Society
Ralph C. d'Arge, University of Wyoming
John W. Berg, The Colorado Regional Cancer Center
Russell F. Christman, University of North Carolina
Cyril L. Comar, Electric Power Research Institute
Eville Gorham, University of Minnesota
Robert C. Harriss, Florida State University
Delbert D. Hemphill, University of Missouri
Margaret Hitchcock, Yale University
Robert J. Moolenaar, Dow Chemical Company
Jean M. Morris, E.I. duPont de Nemours & Company
Peter N. Magee, Fels Research Institute

Ex Officio Members (Panel Chairmen):

Julian B. Andelman, Panel on Low Molecular Weight
    Halogenated Hydrocarbons, University of Pittsburgh
Patrick L. Brezonik, Panel on Nitrates,
    University of Florida
Frank M. D'Itri, Panel on Mercury, Michigan State University
Robert J. Livingston, Panel on
    Kepone/Mirex/Hexachlorocyclopentadiene,
    Florida State University

Staff:

Edward Groth III, Environmental Studies Board
Adele L. King, Environmental Studies Board
Charles R. Malone, Environmental Studies Board (June 1975 -
    December 1976)

ENVIRONMENTAL STUDIES BOARD

John H. Cantlon (Chairman), Michigan State University
Gilbert F. White* (Chairman), University of Colorado
Alfred M. Beeton, University of Michigan
John R. Borchert, University of Minnesota
Jonathan N. Brownell, Paterson, Gibson, Nobel, and Brownell, Montpelier, Vermont
Henry P. Caulfield, Colorado State University
William J. Coppoc*, Texaco, Inc., Beacon, New York
Ralph C. d'Arge, University of Wyoming
Robert Dorfman*, Harvard University
James A. Fay, Massachusetts Institute of Technology
Sheldon K. Friedlander, California Institute of Technology
Estella B. Leopold, University of Washington
Joseph T. Ling, 3M Company, St. Paul, Minnesota
Perry L. McCarty, Stanford University
Dorn C. McGrath*, Jr., The George Washington University
Norton Nelson*, New York University Medical Center
Vaun A. Newill, Exxon Research and Engineering Company, Linden, New Jersey
David Reichle*, Oak Ridge National Laboratories
Richard H. Wellman, Boyce-Thompson Institute, Yonkers, New York
M. Gordon Wolman*, The Johns Hopkins University

Raphael Kasper, Executive Secretary (term beginning October 1977)
Theodore M. Schad, Executive Secretary (term ending September 1977)

*The terms of these Board members expired in September 1977.

CONTENTS

PREFACE		ix
CHAPTER 1	SUMMARY AND RECOMMENDATIONS	1
	Human and Ecosystem Exposures and Effects	1
	Exposures	2
	Effects	3
	Sources, Concentrations, and Global Mass Balances	5
	Sources	5
	Concentrations in the Environment	7
	Global Mass Balances	9
	Control Options	10
	Economic Analysis of Benefits and Costs Associated with Control Measures	12
	Research Recommendations	13
	Human Health	13
	Environmental Occurrence	14
	Analytical Methods	15
CHAPTER 2	INTRODUCTION	17
	References	21
CHAPTER 3	PROPERTIES, SYNTHESIS, REACTIONS, AND ANALYSIS	22
	Properties	22
	Synthesis	28
	Chlorinated Methanes	28
	Brominated Methanes	32
	Iodinated Methanes	32
	Reactions	33
	Air	33
	Water	36
	Analysis	41
	Analytical Techniques	41
	Sampling and Contamination	47
	Calibration	49
	References	51
CHAPTER 4	PRODUCTION AND USES	56
	Production of Halomethanes	56
	U.S. Production of Chloromethanes	56
	World Production of Chloromethanes	64
	U.S. Chloromethane Production Locations and Capacities	64
	U.S. Production of Bromo-, Iodo-, and Mixed Halomethanes	68

	Uses of Halomethanes	68
	Carbon Tetrachloride	68
	Chloroform	72
	Methylene Chloride	72
	Methyl Chloride	72
	Methyl Bromide	72
	Other Halomethanes	73
	References	74
CHAPTER 5	ENVIRONMENTAL OCCURRENCE, TRANSPORT, AND FATE	76
	Sources of Halomethanes to the Environment	76
	Emissions from Product Manufacture and Use	76
	Secondary Emissions and Secondary Formation from Anthropogenic Sources	82
	Summary of Emissions from Anthropogenic Sources	89
	Natural Sources	97
	Environmental Concentrations	100
	Marine Water	100
	Precipitation and Fresh Water	102
	Finished Drinking Water	103
	Indoor Atmosphere	110
	Outdoor Atmosphere	117
	Food	121
	Analysis of Global Mass Balances	131
	The Monohalomethanes and Methylene Chloride	131
	Chloroform	134
	Carbon Tetrachloride	137
	References	146
CHAPTER 6	EFFECTS	154
	Direct Ecosystem Exposure and Effects	154
	Effects on the Stratosphere and Subsequent Biological Impact	156
	The Influence of Halomethanes on Ozone	156
	Climatic Effects	158
	The Impacts	159
	Exposure and Uptake of Halomethanes by Humans	163
	Relative Source Strengths	174
	Confirming Exposures	182
	Methyl Chloride Exposure from Cigarette Smoke	183
	Estimate of Human Health Effects at Actual Exposure Levels	184
	General Toxicity	185

	Extrapolation and Carcinogenic Risk Estimation	186
	Chloroform and Carbon Tetrachloride Risk Estimation	191
	Epidemiological Investigations of Halomethanes in Drinking Water	195
	References	200
CHAPTER 7	CONTROL TECHNIQUES, OPTIONS, AND COSTS	205
	Carbon Tetrachloride	205
	Human Exposure	205
	Stratospheric Exposure	206
	Control Techniques	206
	Chloroform and Other Trihalomethanes	209
	Human Exposure	209
	Control Techniques	210
	Control of Chloroform in Drinking Water	210
	Other Nonfluorinated Halomethanes	223
	References	225
CHAPTER 8	ECONOMIC ANALYSIS OF SELECTED METHODS FOR REMOVING CHLOROFORM FROM DRINKING WATER	227
	Economic Benefits of Reducing Human Exposure to Toxic Substances	227
	A Perspective	227
	Concepts and Principles	228
	Empirical Estimates of Value of Reducing the Probability of Death	230
	Conclusions	232
	Costs of Removing Chloroform and Other Trihalomethanes from Drinking Water	233
	Benefits of Removing Chloroform from Drinking Water Supplies	235
	Benefit-Cost Analyses for GAC and Aeration Systems	240
	Conclusions	246
	References	249
APPENDIX A	Physicochemical Properties of the Nonfluorinated Halomethanes	251
APPENDIX B	Direct Health Effects of Nonfluorinated Halomethanes	263
APPENDIX C	Biographical Sketches of Panel Members and Panel Advisor	291
APPENDIX D	Units of Weight and Concentration	294

PREFACE

Environmental pollutants are not confined to the air, water, or land environments, but move among the various media. Recognizing this, the U.S. Environmental Protection Agency (EPA) established a Scientific and Technical Assessment Report (STAR) program whose purpose was to provide comprehensive reports on pollutants of concern to EPA because of their potential adverse environmental effects through more than one medium.

In early 1975, EPA approached the National Research Council of the National Academy of Sciences to request a series of comprehensive scientific and technical assessment background documents as part of the STAR series. EPA was anxious that these background documents use an environmental "mass balance" approach, i.e., that they attempt to account for the sources, sinks, and receptors for the pollutant as it moves through the environment. They also were to explore the technology and means for controlling the pollutants and the costs and benefits thereof.

The NRC initially agreed to produce two reports. Within the NRC, responsibility for the study was assigned to the Environmental Studies Board of the Commission on Natural Resources. When it agreed to undertake the study, the Environmental Studies Board identified two distinct objectives. The first was to conduct a series of assessments of specific pollutants, as required by EPA. The second was to draw upon the experience gained from conducting a limited number of such assessments, in order to address the broader problem of how such assessments should be done, and how best to use the limited resources of scientific expertise on environmental issues to meet EPA's expanding need for independent, critical scientific evaluations of pollutants.

A Coordinating Committee for Scientific and Technical Assessments of Environmental Pollutants (STAEP) was appointed to oversee panels that would conduct the assessments of specific pollutants and to make recommendations regarding the best methodology for producing such assessments. The pollutants to be studied were mutually agreed upon by EPA and the NRC; pollutants that posed particularly complicated or difficult problems of assessment, given current scientific and technical knowledge, were chosen. The class of Low Molecular Weight Halogenated Hydrocarbons was selected as the topic for one of these reports. A multidisciplinary panel was appointed to undertake this task and began its work in December 1975.

From December 1975 through April 1977, the Panel held a series of nine working sessions, each one to two days in length, through which it conducted its study. At its first meeting the Panel narrowed its topic to the nonfluorinated halogenated methanes. (A discussion of the criteria for this selection appears in the Introduction to the report.) Throughout the study, Panel members and staff conducted a continuing search for widely-dispersed and newly-emerging information. At their meetings, Panel members presented the available information from each of their fields and, pooling their expertise, discussed the interpretation and implications of this information. The study was characterized by a continual flow of new data and the need for postulation and interpretation.

The importance of determining sources, effects, and control methods for nonfluorinated halomethanes is becoming more clearly recognized, as is the present lack of these data. Even as this study is being concluded, new information is emerging which will provide a better data base for assessing nonfluorinated halomethanes in the environment.

Unfortunately, for any study to become most productive, there comes a time when new information can no longer be incorporated and the volume must be published. This can make a report dated before it is published. In face of this, the authors have pointed to areas in which more data are needed and forthcoming, and recommend periodic reevaluation and reassessment.

The Panel appreciates the contribution and information provided by a wide range of people from universities, government, and industry to this effort. I would like to note the dedication, the efforts at coordinating the various aspects of the study (not a small task in a multidisciplinary study), and the attempt to provide and incorporate the most current scientific and technical information available that the Panel members involved in the preparation of this report gave to its production. I especially wish to commend Mr. David Pierotti who was not an appointed Panel member, but responsibly assisted Dr. Rasmussen, attended all the Panel's meetings and worked as hard as the appointed panel members. In addition I would like to acknowledge the contribution of Mr. Mathew J. Schon, who assisted the Chairman in collecting and evaluating some of the literature, as well as compiling and writing a substantial portion of material, especially that dealing with human exposures, uptakes, and health effects.

I would also like to thank the professional staff of the National Academy of Sciences/National Research Council, especially Adele L. King, Staff Officer to the Panel, for their advice and other substantial efforts in coordinating, assisting, and synergizing the activities of the Panel; also Connie Reges and Barbara Matthews for their secretarial support, David Savage for his tireless and excellent typing and revision of the report on the computer editing system, the Manuscript Processing Unit, and the CNR Editorial Office.

Julian B. Andelman, Chairman
Panel on Low Molecular Weight Halogenated Hydrocarbons

Chloroform, Carbon Tetrachloride, and Other Halomethanes:
An Environmental Assessment

CHAPTER 1

SUMMARY AND RECOMMENDATIONS

This study assesses the scientific and technical information available on a class of potential multimedia environmental pollutants, the nonfluorinated halomethanes. This class of compounds includes the chlorinated, brominated, and iodinated methanes, and a few compounds containing two different halogens, such as bromodichloromethane (see Chapter 3, Table 3.1, for a full listing). Of these compounds, the Panel has considered chloroform and carbon tetrachloride in the greatest detail, primarily because the data available on these chemicals are more extensive than on other halomethanes, and they have been shown to be carcinogenic when administered in high doses to test animals.

As a class, nonfluorinated halomethanes are conveniently grouped from the point of view of chemical structure; in other respects, however, they are markedly dissimilar. Their uses, behavior, fate, and effects are sufficiently varied to preclude any uniform judgments, except perhaps a platitude: they have in common the propensity to be at once quite useful and potentially harmful to man. The findings and conclusions of the study, and the recommendations derived from them are therefore usually specific to particular halomethanes, though they may have implications for the group as a whole. The principal findings and conclusions are reviewed in summary below.

HUMAN AND ECOSYSTEM EXPOSURES AND EFFECTS

A quantitative assessment of human and ecosystem exposures and effects will be found in Chapter 6; the summary here deliberately excludes specific figures because these are tentative, and should not be divorced from the discussion of the assumptions on which they were based and the uncertainties with which they are surrounded.

Exposures

Humans are exposed to nonfluorinated halomethanes by three primary routes: intake in water and other fluids, inhalation, and ingestion in food. Although most of the compounds have been identified in either air, liquids, or food, only carbon tetrachloride and chloroform have been observed or studied extensively in all three media.

For the "reference man"--a person with average intakes of water, air, and food--exposed to average environmental concentrations, total uptake of chloroform by all routes is about three times total uptake of carbon tetrachloride. In these average circumstances, chloroform uptakes via fluids, inhalation, and foods are typically in the ratios 7:2:1; carbon tetrachloride uptakes via these routes are typically in the ratio 2:4:1 (see Chapter 6, section on Exposure and Uptake of Halomethanes by Humans). The Panel's calculations of human uptake assume 100 percent absorption of the chemicals from ingested fluids and foods; for inhalation a range of absorption values is used, measured by the difference between concentrations of the compound in inhaled and exhaled air.

It is not known whether humans absorb these chemicals more efficiently when exposed via one of these routes than via others. At face value, however, these figures indicate that, while none of the three routes of exposure is negligible, fluid intake is generally most important for chloroform and inhalation for carbon tetrachloride.

For certain highly exposed members of the population-- those with high intakes of water and/or those in highly polluted areas--total uptakes of chloroform and carbon tetrachloride via all routes may be as much as 40 to 80 times a more common background level of uptake. When exposures are at their highest, the principal routes are air and water for chloroform and air only for carbon tetrachloride. Under conditions of maximum uptake, uptakes of chloroform and carbon tetrachloride via food may be as high as 6.5 to 7.5 times what might be expected under typical conditions of uptake, and may approach the total typical uptake of each compound via all routes.

Methyl chloride and methylene chloride may occur in relatively high concentrations (over 20 parts per billion) in certain indoor air environments. This is about 10^3 times the ambient outdoor concentrations, but 10^{-3} times the U.S. Occupational Safety and Health Administration's standard for worker exposure to these compounds. Trihalomethanes, formed in drinking water from reactions of organic precursors with chlorine, such as bromoform, bromodichloromethane, and dibromochloromethane have been found in drinking water at concentrations of the order of that of chloroform (see Chapter 5, section on Finished Drinking Water).

Anthropogenic sources raise the naturally-occurring levels of human exposures to some of the nonfluorinated halomethanes in air, water, and food. The relative contribution of anthropogenic and natural sources for these compounds is discussed later in this summary in the section on Sources, Concentrations, and Global Mass Balances.

Significant ecosystem exposures to the nonfluorinated halomethanes are rare; these compounds are quite volatile and do not accumulate in the terrestrial or aquatic environments. In the troposphere the compounds are rapidly diluted to low concentrations, and all but the completely halogenated compounds are degraded in processes involving photochemical reactions.

Effects

These compounds have the potential to affect human health, climate, and natural ecosystems both directly through interaction between the compound and the target, and indirectly as a result of mediating phenomena.

Direct Effects

Human exposure in the United States to chloroform and, to a much smaller extent, carbon tetrachloride in public water supplies, is a matter for concern because both compounds are carcinogenic in test animals and therefore pose at least a potential risk to human health (see Chapter 5, section on Finished Drinking Water; Chapter 6, section on General Toxicity). It should be pointed out that there are no data from the industrial work environment--where exposures to carbon tetrachloride and chloroform have been about five orders of magnitude higher than those experienced by the general public--to indicate that the compounds are carcinogenic to man; nor have direct epidemiological studies of industrially exposed groups been conducted to obtain this data. It is unlikely that indirect epidemiological studies in the general population are sensitive enough to detect human risk at the levels estimated from the animal studies, even should such a risk really exist (see Chapter 6, section on Epidemiological Investigations of Halomethanes in Drinking Water).

Estimates in this report of lifetime cancer risk are made on the assumption that the extrapolations from the animal experiments are valid (see Chapter 6, section on Extrapolation and Carcinogenic Risk Estimation); in view of the considerable uncertainties that still exist, these estimates should be used with caution as a basis for regulatory action. The animal data were obtained by administering the compounds by gavage or orally at concentrations many orders of magnitude higher than those to which humans are exposed in food, water, and air. Extrapolations to low levels were based on a multistage mathematical model which assumes that cancer is single cell in origin. No threshold is assumed, and no allowance is made for dose-dependent changes in the way the organism handles the compound, since no data of this nature were available for incorporation into the model. In addition, new data on environmental concentrations or the toxicology of chloroform and carbon tetrachloride may alter these risk estimates.

Bearing in mind these uncertainties, the evidence indicates that human exposures to chloroform and carbon tetrachloride in air pose similar cancer risks, but that the risks are not high from either source. Some urban air concentrations are unusually high, with a correspondingly increased risk to their populations. Under average (21 micrograms per liter) and high (366 micrograms per liter) exposure conditions, chloroform in water poses a greater cancer risk to humans than carbon tetrachloride. For the average individual, water is the major source of risk from chloroform and air from carbon tetrachloride. The risks from food are lower, though again similar for the two compounds. It must be emphasized that the estimated cancer risks from exposures to these compounds in all media are quite low. (See Chapter 6, section on Chloroform and Carbon Tetrachloride Risk Estimation, and the data base in Chapter 5, section on Environmental Concentrations.)

Other than cancer, chronic human health effects of the nonfluorinated halomethanes at environmental levels either cannot be quantified or have not been demonstrated. All trihalomethanes could, however, be a cause for concern because of their structural similarity to chloroform; the potential mutagenicity or fetotoxicity of some compounds in the class, if confirmed, could also be of concern, primarily in the workplace environment and to a lesser degree in drinking water.

We find that acute human health effects from the nonfluorinated halomethanes at the reported environmental exposure levels are not significant (see Chapter 6, section on General Toxicity and Appendix B).

We find no evidence of any direct effect of these compounds on plants, animals, and ecosystems at the concentrations observed in air and natural waters (see Chapter 6, section on Direct Ecosystem Exposure and

Effects). As a result of spills or liquid or air emissions there are situations where local ambient concentrations are much higher than normal. However, even in such cases there is no evidence of harmful effects, although the possibility of chronic or transient effects cannot be excluded.

Indirect Effects

Some estimation of indirect effects of the nonfluorinated halomethanes is possible. One such effect, particularly of carbon tetrachloride, is decrease in stratospheric ozone. Because of large past emission rates of carbon tetrachloride its estimated current ozone reduction (0.5 percent), in contrast to the estimated ultimate or "steady-state" ozone reduction that will be caused by each partially halogenated chloromethane (0.05 percent), is appreciable. However, emissions of carbon tetrachloride to the atmosphere are stable or decreasing, and since the total carbon tetrachloride atmospheric burden is now about 40 times the known annual injection rate, it is plausible to assume it is near the steady-state and will not grow markedly unless emissions increase. We concur with the conclusions of the 1976 NAS/NRC studies on halocarbons (discussed and referenced in Chapter 6, section on Effects on the Stratosphere and Subsequent Biological Impact) that the effects of current emissions of carbon tetrachloride and other nonfluorinated halomethanes are unlikely to be of major significance for human health and ecosystems. (See Chapter 6, section on The Influence of Halomethanes on Ozone.) Were the effects from ozone reduction significant, they would involve human skin cancers, animal cancers, and effects on terrestrial and aquatic ecosystems.

A second indirect effect is the possible impact on climate of the "greenhouse" effect of the nonfluorinated halomethanes. Again we concur with the evaluations of the 1976 NAS/NRC studies that climatic changes and their impacts from current atmospheric concentrations of these compounds are likely to be negligible (see Chapter 6, section on Effects on Climate).

SOURCES, CONCENTRATIONS, AND GLOBAL MASS BALANCES

Sources

Natural Sources

Information on the formation of halomethanes from natural sources is very scarce. The monohalomethanes-- methyl chloride, methyl bromide, and methyl iodide--are generally believed to be primarily natural in origin, with

the ocean as a major source. Reliable estimates of the
quantities of these compounds formed in the oceans cannot,
however, be made with the information available. Natural
sources have also been proposed for methylene chloride,
chloroform, bromoform, iodoform, and carbon tetrachloride.

Anthropogenic Sources

Emissions from anthropogenic production and use are a
major source of several halomethanes to the environment,
particularly carbon tetrachloride and methylene chloride
(see Chapter 5, sections on Emissions from Product
Manufacture and Use; Analysis of Global Mass Balances). For
carbon tetrachloride, chloroform, methylene chloride, and
methl bromide, dispersive uses are the main routes of entry
into the environment from all anthropogenic sources. These
losses from product manufacture and use are largely by
direct emissions to the atmosphere.

Much of the carbon tetrachloride currently released to
the atmosphere is a consequence of its dispersive use as a
fumigant. Emissions of carbon tetrachloride in the United
States from product manufacture and use were highest during
the 1940s and 1950s, but have declined in recent years and
are now estimated to be about 7 percent of the amount of
carbon tetrachloride manufactured each year by U.S.
industries.

Essentially, all the methylene chloride manufactured
enters the atmosphere as a result of dispersive losses from
its end use as a solvent and degreaser. United States
emissions of methylene chloride from product manufacture and
use are estimated at 84 percent of annual U.S. industrial
production.

Carbon tetrachloride and methylene chloride combined
represent about 90 percent of the cumulative U.S. and "free
world"[1] anthropogenic emissions of chloromethanes. However,
whereas emissions of carbon tetrachloride appear to have
leveled off or decreased in recent years, emissions of
methylene chloride (and anthropogenic sources of methyl
chloride) are probably increasing in proportion to their
production rates (see Chapter 5, section on Emissions from
Product Manufacture and Use).

Current annual U.S. emissions of methyl chloride and
chloroform from product manufacture and use are estimated at
2 and 6 percent of annual industrial production,
respectively.

Secondary Sources

Another important category of anthropogenic emissions is secondary formations of trihalomethanes (e.g., from the use of chlorine in industrial processes and water and wastewater treatment) and secondary emissions from anthropogenic activities (e.g., burning of plastics and combustion of gasoline, tobacco, and plant materials). Although the importance of many of these sources is speculative, being based on preliminary data, further research is warranted to determine their significance. (See Chapter 5, sections on Secondary Emissions and Secondary Formation from Anthropogenic Sources; Analysis of Global Mass Balances.) Some of these potential secondary sources are:

• Some trihalomethanes, including chloroform, are formed in municipal water supply systems and in municipal and industrial wastewater effluents (e.g., pulp and paper mills) as a result of the chemical interaction of applied chlorine and certain organic precursors. It appears that carbon tetrachloride, methylene chloride, and methyl chloride are not formed during chlorination.

• Preliminary data based on one measurement indicate that the elevated atmospheric levels of chloroform in metropolitan areas may be related to the use of chlorinated compounds as a gasoline additive.

• The smoking of tobacco is a major source of methyl chloride in closed environments.

The contribution of halomethanes from secondary sources to the global mass balance is difficult to quantify with available data but, except for trihalomethanes, is thought to be small in comparison with direct emissions and natural sources. Secondary sources may, however, be important routes of human exposures in local situations.

Concentrations in the Environment

Water

Methyl iodide, methyl chloride, chloroform, and carbon tetrachloride have been measured in seawater. In rain and snow, low levels of methyl iodide, chloroform, and carbon tetrachloride have been detected. In raw waters, many halomethanes have been found, but mean concentrations are generally less than 1.0 µg/l. In finished drinking waters, quantitative measurements have been made of methylene chloride, chloroform, carbon tetrachloride, bromoform, bromodichloromethane ($BrCHCl_2$), and dibromochloromethane (Br_2CHCl) (see Chapter 5, section on Environmental Concentrations). In addition, methyl chloride, methyl bromide, methylene bromide, and dichloroiodomethane (Cl_2CHI) have been identified. In general, trihalomethane concentrations in finished waters are higher when surface waters are the source of the raw water and prechlorination is practiced.

Air

Six halomethanes--methyl chloride, methyl bromide, methyl iodide, methylene chloride, chloroform, and carbon tetrachloride--have been detected in the outdoor ambient atmosphere, but only carbon tetrachloride, chloroform, and methyl chloride have undergone extensive quantitative analysis. Methylene chloride, methyl iodide, and methyl bromide have been detected, but their concentrations are low, near the detection limits of analytical measurements.

The general background tropospheric concentration of carbon tetrachloride ranges from 120 to 140 parts per trillion. That of chloroform is from 20 to 40 parts per trillion (see Chapter 5, section on Environmental Concentrations: Outdoor Atmosphere), with higher concentrations in marine air; lower levels are normally found in continental air samples. Methyl chloride concentrations in the range of 500 to 1,500 parts per trillion have been reported. Over urban areas, concentrations of carbon tetrachloride, chloroform and methylene chloride are generally higher than those found over marine or non-urban areas.

In indoor atmospheres, concentration levels of chloromethanes generally reflect outdoor atmospheric levels. However, indoor concentrations can reach relatively high levels where an indoor source is present, for instance, in beauty parlors where aerosols or solvents are used thereby releasing methylene chloride, or in rooms where cigarettes are being smoked and methyl chloride is produced.

Food

On the basis of limited information, it appears that chloroform and carbon tetrachloride are widely distributed in edible fish, marine organisms, water birds, and various foods, with typical ranges of 1 to 30 parts per billion (µg/kg). No significant evidence of bioaccumulation via the food chain to higher tropic levels has been found (see Chapter 5, section on Environmental Concentrations: Food).

The use of fumigant mixtures containing chloroform, carbon tetrachloride, and methyl bromide to treat food products results in very low-level residuals in treated foods. Generally, normal dosages and proper storage, handling, and aeration following fumigation, together with normal processing, reduce residuals to negligible levels before food reaches the consumer (see Chapter 5, section on Environmental Concentrations: Fumigant Residues in Food).

Global Mass Balances

It is as yet impossible to construct accurate global mass balances for the compounds in this study. Carbon tetrachloride and chloroform are the only nonfluorinated halomethanes both produced and released by man in large quantities and for which there is much atmospheric information. Even for these two compounds, however, important questions about atmospheric concentrations remain unresolved.

Only four of the remaining compounds have been detected in the atmosphere. These are the monohalomethanes (methyl chloride, methyl bromide, and methyl iodide) and methylene chloride. Although large natural sources have been postulated for the observed tropospheric concentrations of the monohalomethanes, it is possible that significant amounts of methyl chloride and methyl bromide may result from direct anthropogenic emissions or secondary reactions of anthropogenic precursors. Methylene chloride has been measured only a few times, although it is emitted by man in far greater amounts than any of the other compounds, and may contribute significantly to the total chlorine budget of the atmosphere. The main reasons for the lack of methylene chloride measurements are its relatively short atmospheric lifetime and the difficulty with analytical methods in detecting the compound at the low concentrations likely to be present.

The Panel's evaluation of recent studies and historical emissions inventories indicates that most environmental carbon tetrachloride has direct anthropogenic origins (see this chapter, section on Sources; Chapter 5, section on Analysis of Global Mass Balances). Accurate estimates of values for its tropospheric concentration and atmospheric lifetime are needed; for the former, current estimates range from 60 to 140 parts per trillion (volume per volume), and for the latter from 18 to 100 years. Calibration discrepancies among laboratories measuring the atmospheric distribution of carbon tetrachloride need to be resolved, as do questions about how much the United States has contributed to total cumulative global emissions. Estimates of the latter range from 50 to 90 percent, and this Panel estimates 85 percent as of 1976, with the United States contributing about half of current annual emissions (see Chapter 5, Table 5.7).

The relative contribution to atmospheric concentrations of chloroform from anthropogenic and natural sources has not yet been resolved (see Chapter 5, section on Analysis of Global Mass Balances). Direct, worldwide anthropogenic chloroform emissions to the atmosphere over one year have been estimated in another study (discussed in Chapter 5, section on Analysis of Global Mass Balances) as only 1 to 5 percent of total yearly global emissions.

Chloroform concentrations are higher in marine than continental (non-urban) air samples, and the surface mixed layer of the oceans has nearly the same concentration of chloroform as the troposphere, indicating that the marine environment is a source of chloroform. There appears to be natural production by marine organisms to account for some of the elevated levels of chloroform in marine air and surface ocean waters. There may also be other possible sources, such as direct contamination by urban and industrial runoff into coastal waters. The quantitative contribution from these sources is unknown.

The higher levels of chloroform in urban compared to other continental areas indicates that there are also direct or secondary anthropogenic sources for this compound. It appears possible that a significant portion of the chloroform in the environment originates in secondary formation from sources discussed earlier in this chapter.

CONTROL OPTIONS

This study identifies feasible techniques for controlling chloroform and carbon tetrachloride in air and water, but not in food because the principal sources of these compounds in food have not been identified. Since significant potential adverse effects on health and the environment have not been identified at ambient levels of exposure for the other nonfluorinated halomethanes, control options were not considered for those compounds. The options for control identified follow:

- Changes in water disinfection practice would reduce human exposure to chloroform and, to a lesser extent, carbon tetrachloride via drinking water. Controlled chlorine addition, aeration, and carbon absorption appear to be technically feasible methods of control.

- Locally elevated atmospheric concentrations of chloroform and other halomethanes resulting from chlorination practices at municipal water and wastewater treatment plants and pulp and paper mills could be reduced by modifying these and other water and wastewater treatment practices.

- The phasing out of leaded gasoline would eliminate the elevated atmospheric levels of chloroform in metropolitan areas that may be related to the use of chlorinated compounds as a gasoline additive (see Chapter 7, section on Chloroform and Other Trihalomethanes).

- Discontinuing the use of carbon tetrachloride as a grain fumigant would eliminate about 40 to 50 percent of the direct anthropogenic losses of the compound in the United States (see Chapter 7, section on Carbon Tetrachloride). Other chemicals might be able to replace carbon tetrachloride in this application, but data are insufficient to evaluate their efficacy, the implications for health and the environment, or the costs. Moreover, overseas use of carbon tetrachloride, if continued, is likely to lead to continued exposure of United States residents via worldwide contamination of the atmosphere.

- Discontinuing the production of chlorofluorocarbons F-11 and F-12 could eliminate about 30 percent of direct anthropogenic losses of carbon tetrachloride in the United States (see Chapter 7, section on Carbon Tetrachloride).

- Other possible techniques for reducing the input of carbon tetrachloride to the environment include: reduction in losses from industrial producers and users (assuming continued production of F-11 and F-12), curtailment of other small industrial uses of carbon tetrachloride, and reduction of possible secondary production of carbon tetrachloride from anthropogenic sources. Insufficient information was available, however, to evaluate the feasibility of specific control techniques applicable to these sources for reducing the tropospheric burden of carbon tetrachloride.

There is evidence that under certain local conditions human and environmental exposure to several of the nonfluorinated halomethanes could be 10 to 1,000 times higher than typical ambient exposure levels. Places in the vicinity of industrial installations producing or using the compounds, distribution centers, metropolitan centers, and indoor air have been identified as areas with elevated levels of certain compounds. Additional data are needed to document these exposures before evaluating means of controlling the sources.

ECONOMIC ANALYSIS OF BENEFITS AND COSTS ASSOCIATED WITH CONTROL MEASURES

In this study, a prototype benefit-cost analysis was conducted for two control options for reducing chloroform and other trihalomethanes formed in drinking water as the result of chlorination. It should be emphasized that although this analytical method is recommended, the data base upon which this analysis has been made is being expanded and revised even as this report is being written. The Panel's calculations are therefore subject to revision as new data are reported. It should also be noted that this type of analysis is not recommended as a means of making decisions mechanically, but as a tool for providing decision makers with a clearer understanding of some alternatives and their consequences.

The options, selected for analysis because sufficient data were available to make the exercise useful, were the aeration process, which is promising for small water treatment plants of 5 million gallons per day (MGD) capacity and lower, and the granular activated carbon (GAC) process for plants of 10 to 150 MGD capacity. For the aeration process, total annual per capita treatment costs would be at least $7.67 (in small communities of 5 MGD capacity plants), and for the GAC process, the lowest annual total per capita cost would be $2.66 (in large communities with 150 MGD capacity plants).

The benefits to a population of removing chloroform from drinking water can be estimated by the value of the reduced risk to life and health achieved by this removal. The minimum economic value of reducing risk to life and health can be estimated on the basis of such factors as earnings, productivity, and cost of hospitalization. Better estimates of the economic value actually placed by individuals upon small increments in risk can be derived from studies of economic premiums paid for risky occupations. Because of the sensitivity of the estimates to the assumptions involved, this study uses a range of values from $100,000 to $1,000,000 for the per capita value to society associated with reducing the probability of death by 100 percent, or from $1,000 to $10,000 per individual per 1 percent increment in risk over an average lifetime. On the basis of current information, and recognizing the assumptions involved in both the cost and benefit analyses, economic justification exists for removal of chloroform only in large communities in the United States and only where it is found in relatively high concentrations compared to other sources of potable water, and maximum fluid intake is assumed. However, since both treatment processes (GAC and aeration) may also remove carbon tetrachloride, pesticides, and other hazardous substances, additional research on the benefits of removing other impurities in drinking water might greatly increase opportunities for the reduction of premature deaths and illnesses in the United States by treating drinking water (see Chapter 8, Conclusions section).

Because of high costs of treatment and relatively low risks of exposure, removal of chloroform was found to be economically justifiable only in a few cases. Assuming most probable risk to human health (based on extrapolation from high-dose animal feeding studies), economic justification was found only in the case of maximum concentration in drinking water (366 micrograms per liter) and maximum intake (3.7 liters per day) where an estimate of $1,000,000 was used for the value of avoiding a cancer death, and treatment processes could be operated at maximum economies of scale. If a safety factor is allowed by assuming an upper limit risk of five times that of the most probable risk, however, the cost of chloroform removal can be economically justified at low assumed values of avoiding risk, where high initial concentrations are found in drinking water (see Chapter 8, section on Benefit-Cost Analyses for GAC and Aeration Systems). For example, with a 150 million gallons per day GAC treatment process, an initial chloroform concentration must be in excess of 273 micrograms per liter ($\mu g/l$) with a minimum assumed value of $300,000 for avoiding a cancer death. Because treatment costs are usually much more expensive for smaller plants, a similar concentration of chloroform would require an assumption of a considerably greater value for avoiding death in order to justify treatment on economic grounds.

RESEARCH RECOMMENDATIONS

The following research recommendations point to the major gaps in the scientific and technical knowledge available for the nonfluorinated halomethanes. The Panel regards each recommendation as important in providing a more complete basis for an assessment of these compounds as environmental pollutants. The recommendations are not arranged in priority order.

Human Health

1. Carcinogenesis dose-response information should be sought for halomethanes that are of special environmental

concern (e.g., carbon tetrachloride and chloroform). Studies should include both oral and inhalation exposures. In conjunction, more reliable data are needed on human uptake of these compounds from breathing low levels in air 24 hours a day. Metabolic and pharmacological studies should also be conducted for an understanding of potential low-dose carcinogenesis, and risk extrapolation models that include explicit consideration of comparative pharmacokinetics and detoxification mechanisms need to be developed.

2. Further studies are needed to evaluate the chronic health effects, such as carcinogenesis, mutagenesis, and teratogenesis, from low-level exposure to halomethanes, particularly those that have been detected in the environment.

3. Metabolic studies should be considered to evaluate better the absorption, decomposition, and toxic mechanisms of the halomethanes. These studies may also provide insight into potential chronic health effects.

4. Results from recent epidemiological studies based on indirect or ecological data have raised a suspicion that halomethanes in drinking water might be positively associated with human cancer rates. Direct epidemiological studies should be conducted on long-term high exposure groups, if identification of such exposure groups is feasible, to determine whether the evidence supports this association.

Environmental Occurrence

5. To define low levels of exposure more accurately and to evaluate potential health effects of worldwide background levels and incremental risks in local situations, more extensive information should be obtained on environmental levels of the halomethanes in drinking water, food, and both the indoor and outdoor atmosphere.

6. More information is needed on sources of halomethanes to the environment in order to identify anthropogenic sources better and to relate them to environmental levels. For example, pulp and paper mills should be monitored for chloroform emissions both to air and water, and accurate data should be gathered on production and emissions of the major halomethanes, especially carbon tetrachloride, in other countries, particularly the U.S.S.R., the People's Republic of China, and Eastern Europe.

7. To quantify the anthropogenic contribution of chloroform and carbon tetrachloride in the atmosphere, their carbon-14 specific activities should be measured. Low specific activities would indicate a human product from synthesis of petroleum components; values near the atmospheric carbon-14 specific activity value would indicate a natural mechanism for their formation. This concept could also be applied to other halocarbons.

8. Halocarbon concentrations in glaciers should be studied for the history of their atmospheric levels. Glaciers exist at nearly every latitude of the earth, and many of them, both polar and alpine, can have their strata dated by a variety of techniques. They may be a sink for halocarbons as a consequence of regional or global scale distillation reactions; moreover, they may indicate anthropogenic inputs if significant changes in halocarbon concentrations are found in deposited strata dating from the beginning of the twentieth century to the present.

9. The ocean should be assessed as either a source or a sink for chloroform and carbon tetrachloride. For this the following information on the two compounds is essential:

- their concentrations in phytoplankton and zooplankton, to ascertain production in marine organisms and biomagnification;

- their solubility in seawater, a measurement needed to calculate fluxes across the air/sea interface;

- their concentrations in waters below the thermocline to ascertain if the deep ocean is a sink for chloroform or carbon tetrachloride; and

- analyses of marine sediments to ascertain if they are produced there or transported there in organic debris.

Analytical Methods

10. Primary and secondary calibration standards for halogenated hydrocarbons must be developed to assure that analytical methods are valid and reproducible within and among laboratories. Strict quality control and frequent interlaboratory comparisons must be an integral part of any halogenated hydrocarbon analysis program.

NOTE

1. Throughout this report, "free world" is used as a shorthand term for the market economies, as distinguished from centrally-planned economies such as China and the U.S.S.R. "Free world" countries are those for which production data is available.

CHAPTER 2

INTRODUCTION

Growing awareness of the incidence and possible acute and chronic health effects of many trace organic chemicals produced by man is reflected in a wide range of federal regulatory activities. The Occupational Safety and Health Act of 1970, the Clean Air Act Amendments of 1970, the Federal Water Pollution Control Act Amendments of 1972, the Safe Drinking Water Act of 1974, the Resource Conservation and Recovery Act of 1976, and the Toxic Substances Control Act of 1976 are all explicitly or implicitly concerned with the regulation and control of toxic organic and other chemicals in air and water to which humans may be exposed.

Promulgation of standards for many toxic organic chemicals has followed the enactment of this legislation. At the same time, gas chromatography-mass spectrometry (GC-MS) and other analytical techniques have developed to the point where many regulatory facilities and research laboratories can identify and quantify extremely low levels of trace organic chemicals in air (at the part per trillion [ppt] level, or one part in 10^{12}); in water (at the part per billion [ppb] level, or one microgram in a liter of water [μg/liter]); and in other media. The improved analytical sensitivity and the relative ease with which gross mixtures of unknown compounds can now be identified have played a central role in increasing society's concern about possible human and ecological hazards created by trace organic chemicals.

This report is concerned with one small group of organic chemicals whose members are of interest because of their important role in chemical technology, their ubiquity, and their known or potential effects on human health. They include the mono-, di-, tri-, and tetra-chlorinated methanes, the comparable brominated and iodinated methanes, and mixed halogenated methanes. (Table 3.1 relates the common names of these compounds to their formulas and synonyms.) The fluorinated methanes were excluded from consideration because they have been under intense scrutiny elsewhere, as reflected in two 1976 reports by the National Academy of Sciences, _Halocarbons, Environmental Effects of_

Chlorofluormethane Release and Halocarbons: Effects on Stratospheric Ozone (NRC 1976a, 1976b). Some of the deliberations and conclusions of those reports relating to nonfluorinated halomethanes will be used here, because such compounds as carbon tetrachloride, chloroform, and methyl chloride are important constituents of the atmosphere and play a role in ozone reduction and infrared absorption with the concomitant effects.

An assessment of human occupational exposures to the compounds was also excluded because it was outside of the purview of the study. Knowledge of this subject is generally advanced, and the Panel used such information where appropriate, particularly in assessing acute toxicity from various levels of exposure. Criteria for occupational exposure for three of the compounds covered in this report have been issued by the National Institute of Occupational Safety and Health (NIOSH): chloroform (U.S. DHEW 1974), carbon tetrachloride (U.S. DHEW 1975), and methylene chloride (U.S. DHEW 1976).

This analysis was undertaken in response to a request by the Environmental Protection Agency to provide a background study that could be used to develop criteria documents and to assist in regulatory decision making. Thus, the intent was to focus on information that would serve these long-range goals, and not to compile an encyclopedic collection of materials. The Panel was asked to select for study specific shortchain, halogenated hydrocarbons, especially those known or suspected as carcinogens. Each compound was to be assessed in the following five areas:

1. Its chemical and physical properties;

2. Exposures and effects on humans and their environment, with emphasis on dose-response data;

3. Analytical and monitoring techniques;

4. Technology and means for control and regulation; and

5. Costs and benefits of effects and control.

In addition, the Panel was asked to take a multimedia approach in assessing the movement and effects of these compounds and to analyze their global mass balances.

Seven criteria were used to select compounds for study, though every nonfluorinated halogenated methane that was selected does not meet all of these criteria. The criteria were:

- Carcinogenicity

    Although the nonoccupational exposures of humans to the halogenated hydrocarbons are almost always at concentrations not likely to cause acute effects, some of these compounds are either known or likely to have chronic health effects. Carcinogenesis is an important effect of this kind.

- Imminence of Regulatory Action

    Compounds subject to imminent regulation were not evaluated because such evaluations had been or were being made elsewhere. On the other hand, the Panel decided it would not be useful to consider chemicals that might be of regulatory concern only in the distant future.

- Extent of Production, Use, and Human Exposure

    The Panel selected compounds that are produced industrially, used in reasonably large quantities, and are likely to result in measurable human exposures. Thus the compounds are both useful and possibly harmful to humans. By comparing the benefits and any harmful effects of these chemicals, the Panel hoped to make its evaluations in a way that might be useful for assessing other groups of chemicals.

- Quantity and Availability of Information

    The Panel selected compounds for which sufficient useful information was available, but avoided those for which information is so well developed that it has been fully digested and requires no further evaluation.

- Effectiveness of Possible Control Measures

    The likelihood of options for the treatment, control, and regulation of the compounds selected was considered to be important so that alternative control strategies could be considered.

- Existence of Non-anthropogenic Sources

    The possible existence of natural sources of the compounds was a criterion chosen to add an important dimension to the study. This meant that the evaluation of possible effects and of control requirements would have to be made despite any uncertainties that minimum background concentrations could not be affected by human actions.

- Environmental Persistence, Dispersion, and Bioaccumulation

    It was thought desirable to select compounds that reside in the atmosphere and hydrosphere for relatively long periods. Their persistence would be likely to result in bioaccumulation, in movement away from local sources, and, possibly, in dispersion on a global scale with widespread effects on the environment.

Two compounds in this report, carbon tetrachloride and chloroform, have been found to be carcinogenic in animals and meet most of the other study criteria. Several other compounds, which are important in more restricted ways, were not treated as thoroughly. Methyl chloride, bromide, and iodide are of lesser concern because they are thought to be primarily of natural origin, although methyl chloride and methyl bromide may have significant anthropogenic sources and are thus present at higher concentrations in some places. Methylene chloride is currently of interest because of its relatively large dispersive losses, though its short atmospheric lifetime somewhat lessens the need for concern. The mixed bromochloro trihalogenated methanes, for example, are of interest principally because of their formation in the chlorination of drinking water, and probably in the chlorination of wastewater and the bleaching of wood pulp. The other halomethanes are of relatively little concern. The known emissions of the others are small, and in most cases there is little or no available information concerning possible health effects.

REFERENCES

National Research Council (1976a) Halocarbons: Environmental Effects of Chlorofluoromethane Release. Committee on Impacts of Stratospheric Change. Washington, D.C.: National Academy of Sciences.

National Research Council (1976b) Halocarbons: Effects on Stratospheric Ozone. Panel on Atmospheric Chemistry of the Committee on Impacts of Stratospheric Change. Washington, D.C.: National Academy of Sciences.

U.S. Department of Health, Education, and Welfare (1974) NIOSH Criteria for a Recommended Standard. Occupational Exposure to Chloroform. HEW Publication No. (NIOSH) 75-114. Washington, D.C.: U.S. Government Printing Office.

U.S. Department of Health, Education, and Welfare (1975) NIOSH Criteria for a Recommended Standard. Occupational Exposure to Carbon Tetrachloride. HEW Publication No. (NIOSH) 76-133. Washington, D.C.: U.S. Government Printing Office.

U.S. Department of Health, Education, and Welfare (1976) NIOSH Criteria for a Recommended Standard. Occupational Exposure to Methylene Chloride. HEW Publication No. (NIOSH) 76-138. Washington, D.C.: U.S. Government Printing Office.

CHAPTER 3

PROPERTIES, SYNTHESIS, REACTIONS, AND ANALYSIS

The compounds studied in this report are shown in Table 3.1. The first column of the table lists the common name of each compound and the next lists its formula, which are used interchangeably throughout the report. The last column contains frequently encountered synonyms.

PROPERTIES

This section is a discussion of those physical and chemical properties of the nonfluorinated halomethanes that are most important to the compounds' environmental action. A detailed tabulation is given in Appendix A.

At normal ambient temperatures and pressures, most of the nonfluorinated halomethanes are liquids. The exceptions are carbon tetrabromide and iodoform, which are solids, and methyl chloride and methyl bromide, which are gases. However, the boiling points of all the liquids are relatively low, with resulting high vapor pressures. High vapor pressure is an important characteristic of the nonfluorinated halomethanes as a class and has implications for their movement into the atmosphere from manufacturing processes and uses, liquid wastes and spills, and natural water systems.

Figure 3.1 shows the vapor pressures of the chloromethanes at various temperatures. The vapor pressure decreases with the number of halogen atoms of a given type. Thus, at any one temperature the vapor pressures for methyl chloride, methylene chloride, chloroform, and carbon tetrachloride decrease in that order. For a given number of halogen atoms on the methane, the vapor pressure decreases in this order: chlorine, bromine, and iodine. For example, methylene chloride, methylene bromide, and methylene iodide boil at increasingly higher temperatures (40, 97, and 182°C, respectively) at one atmosphere. None of the nonfluorinated halomethanes have boiling points above 218°C.

TABLE 3.1 Nonfluorinated Halomethanes

Common Name (Used in This Report)	Formula	Synonyms
Chlorinated Methanes		
Methyl chloride	CH_3Cl	Chloromethane, monochloromethane
Methylene chloride	CH_2Cl_2	Dichloromethane, methane dichloride, methylene dichloride, methylene bichloride
Chloroform	$CHCl_3$	Trichloromethane, methane trichloride, methyl trichloride, methenyl trichloride, trichloroform, "formyl trichloride" (improper)
Carbon tetrachloride	CCl_4	Tetrachloromethane, methane tetrachloride, perchloromethane, benzinoform
Brominated Methanes		
Methyl bromide	CH_3Br	Bromomethane, monobromomethane
Methylene bromide	CH_2Br_2	Dibromomethane, methylene dibromide
Bromoform	$CHBr_3$	Tribromomethane, methenyl tribromide
Carbon tetrabromide	CBr_4	Tetrabromomethane, carbon bromide
Iodinated Methanes		
Methyl iodide	CH_3I	Iodomethane
Methylene iodide	CH_2I_2	Diiodomethane
Iodoform	CHI_3	Triiodomethane
Carbon tetraiodide	CI_4	Tetraiodomethane
Mixed Halogenated Methanes (selected)		
Bromodichloromethane	$BrCHCl_2$	—
Dibromochloromethane	Br_2CHCl	—
Dichloroiodomethane	CL_2CHI	—

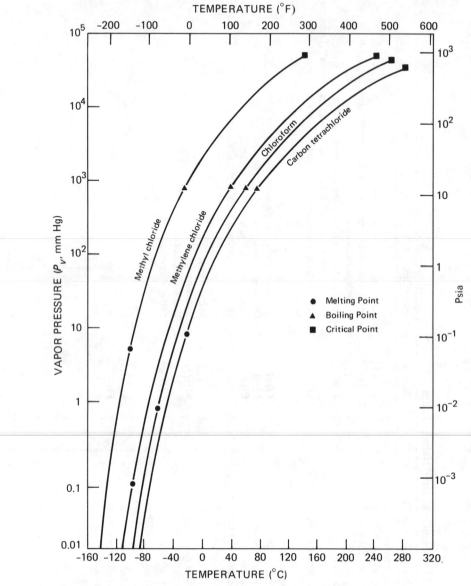

SOURCE: Yaws (1976) Excerpted by special permission from *Chemical Engineering* 83(14):81-89. Copyright 1976 by McGraw-Hill, Inc., New York, N.Y. 10020.

FIGURE 3.1 Vapor pressures of chloromethanes at various temperatures.

The liquid densities of the halomethanes, most of which are liquids under typical ambient conditions, are particularly important in determining their behavior and fate in spills to natural waters. As typified by the chloromethanes, shown in Figure 3.2, their liquid densities are significantly greater than that of water (methyl chloride is a gas at normal pressures and temperatures). The densities of comparable bromides and iodides are even higher. It can be expected that, after a liquid spill of these compounds, they will tend to settle in any receiving water before dispersion, volatilization, emulsification, or solubilization takes place, except of course for methyl chloride.

In all cases, the specific gravities of the halogenated methane vapors are greater than that of air because their molecular weights are greater than those of nitrogen and oxygen, the principal air constituents. This property is of interest primarily when a large gaseous emission occurs. In these instances, such as when a tank ruptures and there is a gaseous emission, the plume will tend to settle to the ground before there is any mixing with air. In time, however, the gas or vapor will be incorporated into the ambient air.

Solubility in water is an important property in assessing the fate and movement of these compounds in the environment, especially between the hydrosphere and the atmosphere. When solubility in water is listed for a halomethane that normally is a liquid (see Appendix A), the datum refers to the aqueous saturation concentration in contact with the liquid at that temperature and at one atmosphere of pressure. When the compound normally is a gas or vapor at the cited temperature, the solubility in water is that in equilibrium with the gas or vapor at one atmosphere. For any given series of liquid halogen compounds, solubility decreases with an increasing number of halogen atoms, just as does the vapor pressure. Thus, solubilities decrease in the order of methylene chloride, chloroform, and carbon tetrachloride. Similarly, the comparable chlorides generally are more soluble than the bromides, which are more soluble than the iodides. Temperature has little influence on solubility of these compounds where its effect is known, but other solutes in water, such as inorganic ions, will probably affect the solubility of the nonfluorinated halomethanes through the well-known "salting-out" or "salting-in" mechanisms, although data for such effects generally are not available for these compounds.

Appendix A contains tabulated information on the partition coefficient for the chlorinated methanes. As the aqueous solubilities decrease with an increasing number of

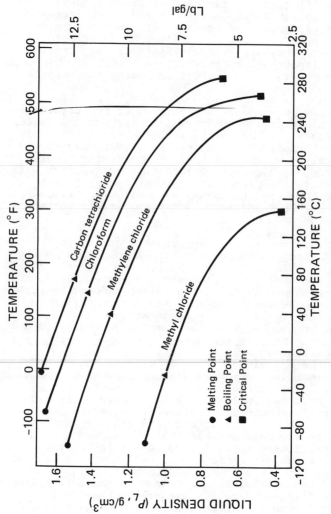

FIGURE 3.2 Liquid density of chloromethanes at various temperatures.

halogen atoms, the partition coefficients increase. The partition coefficient is useful in assessing the direction of any deviation from equilibrium when both water and ambient air concentrations are known. For example, in the continental United States a typical urban air concentration for chloroform is 0.2 ppb (v/v), or 1×10^{-3} µg/l at 25°C. Using a partition coefficient of 0.16, this indicates that $(1/0.16) \times 1 \times 10^{-3}$, or 0.6×10^{-2} µg/l (ppb), is the chloroform concentration in water that would be in equilibrium with the specified air concentration. In fact, although many reported values for marine waters containing chloroform fall in this concentration range, most fresh waters have considerably higher values, with concentrations of at least 1 µg/l. For such fresh water and air concentrations, the equilibrium is widely displaced to the point that chloroform should evaporate from water to air.

Some measurements of evaporation rates for the chlorinated methanes in dilute aqueous solution have been made by Dilling et al. (1975); the results are tabulated in Appendix A. Although various conditions, including water volume, temperature, geometry, and stirring rate affect the evaporation rates it was found that evaporation rates did not vary greatly with the specific compound, and that 50 percent depletion occurred in 20 to 25 minutes.

Another heterogeneous interaction that can affect the environmental movement of halomethanes is their possible sorption onto solid surfaces from an aqueous solution. This could retard evaporation or, if the sorption is particularly strong, natural sorbents could serve as sinks for these compounds. This phenomenon also can serve in treatment processes as, for example, in the use of activated carbon as a sorbent to remove halomethanes from aqueous solutions and gaseous systems.

There are few measurements of halomethanes sorbed onto natural materials, but some indications of the sorption of these compounds from aqueous solutions onto natural materials can be found in the experiments of Dilling et al. (1975). These investigators determined the effect of such sorption on the rate of evaporation of halomethanes. They found that bentonite clay sorbed relatively large quantities (20 percent of 1.0 ppm initial concentration) of these compounds after 20 minutes and that peat moss in water sorbed the compounds and thereby affected their evaporation rates.

Other measurements of halomethanes sorbed onto natural materials were made by Pearson and McConnell (1975), who found carbon tetrachloride and chloroform in marine bay sediments at concentrations in the range of a few micrograms per kilogram. On the other hand, Neely et al. (1976)

concluded, in postulating a model for the change in chloroform concentration in a river after a spill, that their model precluded the existence of a dynamic equilibrium between the chloroform and river bottom mud.

SYNTHESIS

Chlorinated Methanes

The chloromethanes—methyl chloride, methylene chloride, chloroform, and carbon tetrachloride—usually are manufactured by the direct chlorination of methane in the presence of ultraviolet light (uv) or a catalyst. Chlorine (Cl_2) and methane (CH_4) react to produce a mixture of all four chloromethanes. The reactions are:

$$CH_4 + Cl_2 \longrightarrow CH_3Cl + HCl$$

$$CH_3Cl + Cl_2 \longrightarrow CH_2Cl_2 + HCl$$

$$CH_2Cl_2 + Cl_2 \longrightarrow CHCl_3 + HCl$$

$$CHCl_3 + Cl_2 \longrightarrow CCl_4 + HCl$$

A typical chlorinated methane yield is 99-100 percent based on chlorine, or 85-90 percent based on methane (Faith et al. 1965).

The reaction conditions can be controlled and modified to produce various percentage yields of the four chloromethanes depending upon the products desired. Recycling the less chlorinated compounds (CH_3Cl and CH_2Cl_2) along with more Cl_2 to additional reactors results in larger yields of chloroform and carbon tetrachloride. The product mixture is separated from unreacted materials by scrubbing with a refrigerated mixture of liquid chloromethanes. The products are purified by washes of hot water, alkali, and concentrated sulfuric acid and the final separation is accomplished by fractional distillation. Figure 3.3 is a schematic flow chart of the production of the four chloromethanes by the chlorination of methane.

Methyl chloride also may be commercially manufactured using the action of hydrochloric acid (HCl) on methyl alcohol (CH_3OH) with the aid of catalysts in either the liquid or vapor phase. The reaction is:

$$CH_3OH + HCl \xrightarrow{\text{catalyst}} CH_3Cl + H_2O$$

The crude product is purified and separated as shown on the flow chart for the chlorination of methane. The vapor phase

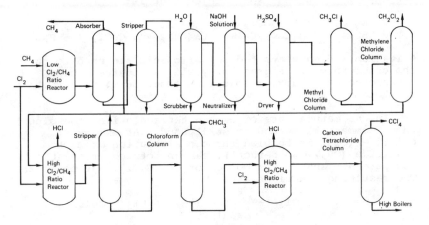

SOURCE: Austin, G. T. (1974) Industrially significant organic chemicals, Part 3. *Chemical Engineering* 81(16):87-92.

FIGURE 3.3 Schematic flow chart of production of chloromethanes by chlorination of methane.

modification of this reaction results in a yield of about 95 percent based on methyl alcohol; it is the most economical process for producing methyl chloride alone.

Methylene chloride also may be synthesized by the reduction of chloroform ($CHCl_3$) by trisilane (Si_3H_8) in the presence of zinc and acetic acid. The reaction is:

$$4\ CHCl_3 + Si_3H_8 \xrightarrow{Zn} 4\ CH_2Cl_2 + Si_3H_4Cl_4$$

However, this reaction is not now economical.

Another major industrial synthesis of chloroform uses the action of bleaching powder ($CaOCl_2 \cdot H_2O$) on acetone (CH_3COCH_3), acetaldehyde (CH_3CHO), or ethyl alcohol (CH_3CH_2OH). The reaction with acetone is:

$$2\ CH_3COCH_3 + 6\ CaOCl_2 \cdot H_2O \longrightarrow$$
$$2\ CHCl_3 + (CH_3COO)_2Ca + 2\ Ca(OH)_2 + 3\ CaCl_2 + 6\ H_2O$$

Agitation and concentrated sulfuric acid are used to purify the chloroform. The greatest yield is obtained using acetone as the charging solution, about 86-91 percent based on acetone.

In the manufacture of carbon tetrachloride, the oldest known commercial method is still used. An excess of chlorine (Cl_2) is bubbled through a solution of carbon disulfide (CS_2, ~40 percent), carbon tetrachloride (CCl_4, ~50 percent) and sulfur monochloride (S_2Cl_2, ~10 percent). The resulting reactions are:

$$CS_2 + 3\ Cl_2 \xrightarrow[\text{catalyst}]{Fe} S_2Cl_2 + CCl_4$$

$$CS_2 + 2S_2Cl_2 \longrightarrow 6\ S + CCl_4$$

This process has the advantages of producing no by-products other than reusable sulfur (used to regenerate carbon disulfide, $2\ S + C \longrightarrow CS_2$) and the separation of the product is easy. The carbon tetrachloride yield is about 90 percent based on carbon disulfide. Figure 3.4 is a schematic flow chart of carbon tetrachloride production from carbon disulfide and chlorine.

Carbon tetrachloride is also produced commercially by reaction of propylene with Cl_2. Products are CCl_4 and $CCl_2=CCl_2$. The ratio of CCl_4 to $CCl_2=CCl_2$ can be varied within limits to meet market needs.

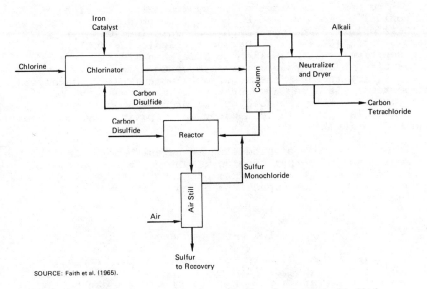

SOURCE: Faith et al. (1965).

FIGURE 3.4 Schematic flow chart for the production of carbon tetrachloride from carbon disulfide and chlorine.

Brominated Methanes

Methyl bromide is commonly produced by the interaction of methyl alcohol (CH_3OH) and hydrobromic acid (HBr) in a reactor. Distillation through a reflux column separates methyl bromide from the reactant mixture. The crude product is purified by low-temperature fractional distillation.

Another method of synthesis is to add sulfuric acid (H_2SO_4) to a strong sodium bromide (NaBr) and methyl alcohol solution. Methyl bromide is separated and purified in the same manner.

The most common method of synthesizing methylene bromide, bromoform, and carbon tetrabromide is to treat their chlorine analogs--methylene chloride (CH_2Cl_2), chloroform ($CHCl_3$), and carbon tetrachloride (CCl_4), respectively--with a slightly more than stoichiometric quantity of anhydrous aluminum bromide or with hydrobromic acid (HBr) in the presence of an aluminum halide catalyst. The result is a replacement-type reaction whereby the chloride atom is replaced by the bromide atom. The crude product is washed in cold water and separated by fractional distillation.

Other methods of producing methylene bromide involve the bromination of methyl bromide (CH_3Br) or the reduction of bromoform ($CHBr_3$) with sodium arsenite ($NaAsO_2$) in an alkali solution, but these reactions do not have commercial significance.

Iodinated Methanes

Methyl iodide usually is prepared by the interaction of methyl alcohol (CH_3OH) with red phosphorous and iodine (I_2). However, there are a number of other ways it can be prepared. The interaction of dimethyl sulfate and sodium or potassium iodide in an aqueous solution will produce methyl iodide as will the reaction of dimethyl sulfate with an aqueous iodine slurry containing a reducing agent, such as powdered iron or sodium bisulfite.

Other methods of synthesis for methyl iodide include the interaction of methyl alcohol with hydroiodic acid (HI) and the high-yield reaction of methyl alcohol, iodine, and borane (B_2H_6). The interaction of methyl alcohol, sodium iodide, and sulfuric acid (H_2SO_4) also will produce methyl iodide. Purification of methyl iodide is accomplished by fractional distillation.

The most convenient way to produce methylene iodide is to treat iodoform (CHI_3) with sodium arsenite (Na_3AsO_3) and sodium hydroxide (NaOH). The iodoform undergoes a reduction reaction to form methylene iodide. The reaction is:

$$CHI_3 + Na_3AsO_3 + NaOH \longrightarrow CH_2I_2 + NaI + Na_3AsO_4$$

Other methods of producing methylene iodide involve the oxidation of iodoacetic acid (ICH_2CO_2H) by potassium persulfate ($K_2S_2O_8$) and the action of iodine (I_2), sodium ethoxide, and hydroiodic acid on iodoform (CHI_3).

Iodoform is prepared industrially by the electrolysis of acetone (CH_3COCH_3) or alcohol solutions of a mineral iodide in the presence of sodium carbonate (Na_2CO_3). The Kirk-Othmer Encyclopedia of Chemical Technology (Standen 1964) notes that treating any compound containing the group CH_3CO linked to hydrogen or carbon, or any structure that may be oxidized to this group with an alkali and iodine mixture, will produce iodoform. Iodoform also may be prepared with an excellent yield by the redistribution reaction of chloroform ($CHCl_3$) and methyl iodide.

REACTIONS

Air

The major atmospheric reactions of the six important atmospheric nonfluorinated halomethanes are listed in Table 3.2, along with the best current data on their reaction rate constants. Some reactions that may be significant (i.e., CH_3I + OH, CH_2Cl_2 + hν, and $CHCl_3$ + hν) are not listed because there are no data available on their rate constants. However, using the available rate constants and the best estimates of OH and singlet or excited oxygen atom ($O(^1D)$) concentrations in the atmosphere, approximate atmospheric lifetimes for these six nonfluorinated halomethanes can be calculated.

The three major reactions, OH radical abstraction, reaction with $O(^1D)$ and ultraviolet photolysis, are of varying importance for the different compounds. For the halocarbons containing carbon-hydrogen bonds OH radical abstraction of hydrogen atoms is generally the predominant reaction in the troposphere, although methyl iodide has such a large ultraviolet absorption in the troposphere that uv photolysis is the major sink for this compound. For carbon tetrachloride, which has no C-H bonds, OH radical abstraction is negligible, and the major reaction pathways are uv photolysis and reaction with $O(^1D)$ atoms. The tropospheric lifetime of CCl_4 for OH radical attack is greater than 100 years.

TABLE 3.2 Major Atmospheric Reactions of Significant Low-Molecular-Weight Halogenated Hydrocarbons

Reactions		Rate Constant[a]		Reference
Methyl Chloride (CH_3Cl)	$CH_3Cl + h\nu$[b]	1×10^{-11} 1×10^{-8} 1×10^{-7} 2×10^{-7}	(20 km) (30 km) (40 km) (50 km)	Robbins (1976)
	$CH_3Cl + OH$	$1.69 \times 10^{-12} \exp(-1066/T)$[c] 8.5×10^{-14}		Davis et al. (1975)[1] Cox et al. (1976)
Methylene Chloride (CH_2Cl_2)	$CH_2Cl_2 + OH$	$(1.16 \pm 0.12) \times 10^{-13}$ 1.04×10^{-13}		Davis (1975)[2] Cox et al. (1976)
Chloroform ($CHCl_3$)	$CHCl_3 + OH$	$1.0 \times 10^{-12} \exp(-630/T)$ 1.68×10^{-13} $(1.10 \pm 0.15) \times 10^{-13}$ 1.01×10^{-13}		Howard and Evenson (1975)[3] Cox et al. (1976) Davis (1975)[2] Howard (1975)[4]
Carbon Tetrachloride (CCl_4)	$CCl_4 + h\nu$	9×10^{-9} 9×10^{-8} 5×10^{-7} 5×10^{-6}	(20 km) (25 km) (30 km) (40 km)	Molina and Rowland (1974)
	$CCl_4 + O(^1D)$	$\lambda \geqslant 4.2 \times 10^{-10}$		Atkinson et al. (1976)
	$CCl_4 + OH$	$<4 \times 10^{-15}$ $<1 \times 10^{-16}$		Howard (1975)[4] Cox et al. (1976)
Methyl Bromide (CH^3Br)	$CH_3Br + h\nu$	1×10^{-11} 1×10^{-9} 1×10^{-6} 5×10^{-5}	(16 km) (20 km) (30 km) (50 km)	Robbins (1976)
	$CH_3Br + OH$	$2 \times 10^{-12} \exp(-1200/T)$ $8.3 \times 10^{-13} \exp(-916/T)$ 3.5×10^{-14}		Davis et al. (1975)[1] Davis (1975)[2] Howard (1975)[4]
Methyl Iodide (CH_3Cl)	$CH_3I + h\nu$	6×10^{-6}		Yung et al. (1975)

[a] Bimolecular rate constants in cm^3 molecule^{-1} sec^{-1}; unimolecular rate constant in sec^{-1}.
[b] $h\nu$ is the energy of the quantum of light associated with the particular frequency ν.
[c] T is absolute temperature.

NOTE: Complete references to footnotes are listed in Chapter 3, Notes.

The most important sink for methyl chloride atoms in the atmosphere is OH radical attack in the troposphere. Estimates of the atmospheric lifetime of methyl chloride based upon different OH profiles in the troposphere range from 0.37 years (Cox et al. 1976) to 0.9 years (Yung et al. 1975). Thus methyl chloride has a reasonably long atmospheric lifetime, long enough to allow considerable mixing throughout the troposphere. A study by Spence et al. (1976) showed that the major reaction product of atmospheric oxidation of methyl chloride is formyl chloride (HCOCl). Because they used uv light and initiated their reactions with chlorine, these results may not be valid as indications of reaction rates in the troposphere; however, the major oxidation products should be the same.

The situation is similar for methylene chloride (CH_2Cl_2) and chloroform ($CHCl_3$), although both these compounds are more reactive than methyl chloride to OH attack. Cox et al. estimate atmospheric lifetimes of 0.3 years and 0.19 years for CH_2Cl_2 and $CHCl_3$, respectively. Yung et al. estimate the lifetime of chloroform in the atmosphere at 0.32 years. According to Spence et al. (1976), the major oxidation products of CH_2Cl_2 are CO, CO_2, and $COCl_2$ (phosgene), while phosgene is the dominant product of the atmospheric oxidation of chloroform.

As stated above, carbon tetrachloride has an atmospheric lifetime of greater than 100 years for reaction with OH radicals. The major sinks for CCl_4 in the atmosphere are uv photolysis and reaction with O(^{1}D) atoms in the stratosphere. Due to its large uv absorption in the window region near 200 nm, CCl_4 should decompose primarily by photodissociation. The major product of carbon tetrachloride photooxidation is phosgene (Atkinson et al. 1976).

Methyl bromide behaves in a manner similar to methyl chloride, with OH radical attack the most important reaction in the troposphere. However, uv photolysis becomes the dominant loss mechanism for methyl bromide above 25 kilometers, whereas photolysis and OH attack remain comparable in magnitude for CH_3Cl (Robbins 1976). The major products of photodissociation or OH attack upon methyl bromide are presumably CO and CO_2. The atmospheric lifetime of CH_3Br is somewhat shorter than that of CH_3Cl, primarily because of its rapid photolysis above 25 kilometers, but its behavior in the troposphere should be very similar.

Finally, methyl iodide has such a large uv absorption coefficient that uv photolysis is the major sink for CH_3I in the troposphere (Yung et al. 1975). The estimated atmospheric lifetime of methyl iodide is less than two days, which would limit its occurrence in the atmosphere to the vicinity of its production. The major products of the photodissociation of methyl iodide should be CO and CO_2. It has also been suggested that methyl iodide may react with chloride or bromide ions in sea spray to form methyl bromide and methyl chloride.

The most significant oxidation reaction product of the major atmospheric nonfluorinated halomethanes is phosgene, $COCl_2$. Phosgene is a highly toxic substance, but is believed to have an extremely short atmospheric lifetime due to further attack by free radicals and uv photolysis (Spence et al. 1976). However, Singh et al. (1977) recently claimed to have measured phosgene in the atmosphere, at low part per trillion levels, and the results would appear to indicate a somewhat longer atmospheric lifetime than has generally been assumed. Nonetheless, phosgene should have a lifetime of no more than a few days, and therefore it might be locally significant but should not build up in the troposphere, nor is it likely to enter the stratosphere in sufficient amounts to be of importance to the stratospheric chlorine budget.

Water

There is a variety of possible mechanisms for the degradation or transformation of halomethanes in natural water systems, including chemical, photochemical, and biologically mediated reactions. Unfortunately, there is a dearth of specific information about these mechanisms, and it is therefore difficult to estimate the stability and reactivity of the halomethanes in such water systems.

McConnell et al. (1975) concluded there is evidence that some of the one- or two-carbon chlorinated methanes can be metabolized by mammalian tissues, but not by microorganisms. With bacteria from domestic sewage in a conventional biochemical oxygen demand (BOD) test, it has been shown that carbon tetrachloride is not biochemically oxidized in five days (Heukelakian and Rand 1955). However, results of similar experiments with chloroform gave mixed results; in one investigation there was no reaction, while in another 0.008 grams of oxygen were used per gram of chloroform.

The enzymology of carbon-halogen bonds has been reviewed by Goldman (1972). He notes that the carbon-iodine bond is quite labile and is not, therefore, likely to be encountered in natural environments. He also indicates that, although there is little information on enzymatic rupture of the carbon-bromine bond, it seems to behave like the carbon-chlorine bond. Most of his discussion centers around enzymatically-mediated reactions of fluorine and chlorine compounds, which generally contain more than one carbon and in which other molecular structural features, not present in halomethanes, determine reactivity. Some enzymes for these reactions involving the halogen bond have a bacterial origin. However, only mammalian enzymes (from rat liver) are cited as catalyzing the reactions of halogenated methanes. These reactions include the dehalogenation of mono- and di-halogenated methanes.

There are relatively few data relating to similar reactions in aqueous media, although the photochemical reactions of several halomethanes have been studied intensively in gaseous systems. For a year Dilling et al. (1975) periodically analyzed sealed aqueous solutions of chloroform and methylene chloride exposed to natural sunlight and ambient outdoor temperatures (-20 to +40°C). They concluded that the major reactions accounting for transformations of chloroform and methylene chloride were hydrolysis and that the half-lives were nearly the same in the light or dark. Presumably, there was no significant photochemical degradation. However, in these experiments the aqueous solutions were relatively "clean" in that, except for oxygen, they did not contain other compounds, especially organics, that could absorb light more strongly and cause photosensitized reactions with the halomethanes. Thus, in order for light-induced reactions to occur, there probably would have to have been direct photolysis, which is unlikely for molecules that are poor absorbers of visible and near-ultraviolet light.

Wolfe et al. (1976) note that sensitized photolysis reactions, involving the presence of other light-absorbing organics, can occur in natural waters. They also state that singlet oxygen atoms can be generated in water by energy transfer from substances that absorb sunlight. These excited oxygen or peroxide species could then photooxygenate otherwise unreactive organics, like halomethanes.

Mabey and Mill (1976) discussed the kinetics of some of the possible chemical reactions of a variety of organic chemicals in natural waters, including organic halides as a class, and some of the halomethanes. Among the reactions considered was oxidation involving peroxy, alkoxy, and other radicals, as well as singlet molecular oxygen, all photochemically generated. In the case of the reaction of alkoxy radical with various organic C-H bonds at 40°C, they calculated that the half-life was greatest for those species with chlorine on the carbon, compared to several other types, including alkanes, olefins, ethers, and ketones. Although the possibility exists for the photochemically-induced oxidation of the halomethanes in water, these investigators also point out that there is a variety of natural materials in water that can inhibit or retard radical oxidation processes. These include numerous carotenid-like molecules, which can function efficiently as quenchers of singlet molecular oxygen.

Hydrolysis is another type of reaction that could account for transformations of halomethanes in an aqueous solution, as typified by the following reaction for methyl chloride:

$$CH_3Cl + H_2O \longrightarrow CH_3OH + H^+ + Cl^-$$

To the extent that such chemical reactions occur, they can ultimately lead to the complete oxidation of the halomethanes, since these hydroxylated products are likely to be much more susceptible to enzymatic or other reactions.

Dilling et al. (1975) found that in their sealed aqueous systems the degradations of methylene chloride and chloroform had half-lives of 18 and 15 months, respectively. They concluded that hydrolysis was the likely mechanism. A compilation of estimated and measured half-lives for such hydrolysis reactions is shown in Table 3.3.

Although several of the half-lives shown on this table are calculated estimates and must be used with care, there are a few general trends that should be noted. Among the monohalogenated methanes, the rate of hydrolysis probably increases in the sequence chloride, bromide, and iodide. The rate shown for methyl iodide is at a higher temperature, but the carbon-iodide bond is recognized as more labile than the others. This sequence probably applies to those molecules with more halogen atoms as well. Among the chlorinated methanes, the rate of hydrolysis increases in the sequence carbon tetrachloride, chloroform, methylene chloride, and methyl chloride; this sequence is likely to hold for the other halogenated methanes as well.

There may be a pH effect on hydrolysis rates for some halogenated methanes, as is shown for chloroform in Table 3.3. Mabey and Mill (1976) indicate that there can be both acid- and base-catalyzed hydrolysis reactions for these compounds, and that, depending on their relative magnitudes, there can be a substantial effect of pH within the naturally occurring regions of 4 to 10. They attribute the lack of pH effect for methyl chloride and bromide to the absence of a known acid-catalyzed mechanism. In such cases, the base-catalyzed mechanism would result in a significant pH effect only above pH 10. In contrast, Mabey and Mill showed that for chloroform there is a substantial pH effect on hydrolysis, the rate increasing with pH.

It should be pointed out that the Dilling et al. (1975) measured rates of degradation for methylene chloride and chloroform, which are shown on Table 3.3, are presumed to be

TABLE 3.3 Estimated Half-Lives of Halogenated Methanes in Hydrolysis Reactions in Aqueous Solution

Compound	$t_{1/2}$ hours	$t_{1/2}$ years	Temp., °C	pH	Reference
Methyl Chloride	10^4	1	25	4-10	Mabey and Mill (1976)
Methyl Bromide	480	5×10^{-2}	25	4-10	”
Methyl Iodide	8	10^{-3}	69	7	Moelwyn-Hughes (1957)
Methylene Chloride	6×10^6	7×10^2	–	–	Gordon (1976)
	1.5×10^4	1.5	25	–	Dilling et al. (1975)
Chloroform	3×10^8	3×10^4	25	4	Mabey and Mill (1976)
	3×10^7	4×10^3	25	7	”
	3×10^4	4	25	10	”
	1.3×10^4	1.3	25	–	Dilling et al. (1975)
Carbon Tetrachloride	6×10^8	7×10^4	–	1-7	Johns (1976)

due to hydrolysis, and are substantially faster than the estimated values. It is also of interest to note that there is a great range of hydrolysis rates that can be expected for these compounds in natural water systems, as typified by stable carbon tetrachloride at one end of the scale and labile methyl iodide at the other.

Another possible reaction involving halomethanes in aqueous systems is one in which a halide ion in solution may displace a halogen atom, as follows:

$$CH_3Br + I^- \underset{R_2}{\overset{R_1}{\rightleftarrows}} CH_3I + Br^-$$

These reactions are likely to be second order, dependent on concentrations of both the halomethane and the reacting halide (Moelwyn-Hughes 1957). Using available information for the forward reaction rate constant, R_1, at 25°C and assuming a typical iodide content in seawater of 5×10^{-7}M (molar), one can calculate a half-life of about 44 years for the reaction of iodide with methyl bromide. This is longer by three orders of magnitude than that shown in Table 3.3 for the hydrolysis reaction for methyl bromide, but the reverse reaction probably is considerably faster. That is, the bromide ion is more likely to displace the iodide to form methyl bromide because of the typically greater concentrations of bromide compared to iodide in natural water systems, both fresh and seawater, and the greater stability of the carbon-bromine bond.

These considerations make possible some generalizations regarding the stability of halomethanes in aqueous systems. Because the bond energies of the alkyl chlorides decrease in the sequence chloride, bromide, and iodide, the stability of the halomethanes with the same number of halogen atoms also can be expected to decrease in that order. This should be reflected in chemical reactions involving hydrolysis, displacement by other halide ions, and oxidation, as well as in enzymatically-controlled reactions. Furthermore, because the concentrations of reacting species play an important role in reaction rates and equilibria, the direction of displacement reactions involving halide ions in natural waters generally will favor the formation of chlorinated methanes at the expense of the brominated ones, and these in turn will be favored over iodinated methanes. Both bond energies and relative concentrations of the halide ions in natural waters act in concert.

Although the C-Cl bond energies of other alkyl halides are almost identical with those of chloroform and carbon tetrachloride, hydrolysis rates for chlorinated methanes indicate increasing stability with the number of halogen atoms. This may be due to steric effects on the reaction rates. A comparison of the gas phase reaction rate constants for chlorinated methanes reacting with OH radicals also demonstrates this trend, as shown in Table 3.2. Thus, it is likely that sensitized photooxidation reactions in aqueous solution involving excited molecular oxygen or other species would have rate constants following the same sequence.

In summary, the stabilities of halomethanes in aqueous systems are likely to follow those of their half-lives in hydrolysis. The extremes are the poor stability of methyl iodide and the high stability of carbon tetrachloride. Unstable compounds are likely to react rather quickly and not be encountered in natural waters. But the more stable compounds are likely to have sufficiently long half-lives to make possible their ultimate decomposition or degradation only after their movement from natural waters into the atmosphere, provided that no other non-chemical degradation reactions in water occur.

ANALYSIS

Analytical Techniques

Air

There are four practical methods to measure air concentrations of the halogenated hydrocarbons:

- Gas chromatography with an electron-capture detector (EC-GC);

- Gas chromatography-mass spectrometry (GC-MS);

- Long-path infrared absorption spectroscopy, usually with preconcentration of whole air and then separation of the compounds by gas chromatography (GC-IR); and

- Infrared solar spectroscopy, using the solar spectrum at large zenith angles to obtain great path lengths through the atmosphere.

Each method has advantages and disadvantages and applications for which it is best suited. A major drawback of all these techniques is that they do not allow real-time continuous measurements of the halocarbons at ambient levels in the environment.

Gas chromatography with an electron capture detector (EC-GC) is by far the most widely used method to measure environmental levels of halogenated hydrocarbons. Most atmospheric values reported for these compounds have come from EC-GC analyses, starting with Lovelock's work in the early 1970s and continuing up to the present time (Krey et al. 1976; Lillian et al. 1975; Lovelock 1975; Lovelock et al. 1973; Montgomery and Conlon 1967; Murray and Riley 1973a, 1973b; Pearson and McConnell 1975; Rasmussen 1976; Singh 1976; Su and Goldberg 1976; Williams 1965). There are a number of reasons for the popularity of the EC-GC method. First, the instruments are commercially available and relatively inexpensive, costing between $5,000 and $10,000. Second, the instruments are rugged and easy to operate; they have been used successfully on board ships and aircraft, and they can be operated easily by relatively unskilled personnel. They also are extremely sensitive for most of the halogenated hydrocarbons and are capable of measuring levels of a few parts per trillion (v/v) for such compounds as CCl_4 in sample volumes of only 5 or 10 ml of ambient air. Finally, with EC-GC it is possible to make quasi-continuous measurements, i.e., measurements every 15 or 20 minutes, which is more often than can be done by other means.

The major disadvantage of electron capture detectors is the fact that they are extremely compound-dependent in their sensitivity. Thus, although CCl_4, $CFCl_3$, CF_2Cl_2, and other halocarbons can be easily detected down to 1 part per trillion (v/v) in the atmosphere with no special effort, compounds such as CH_3Cl require concentration by 2 to 3 orders of magnitude before they can be measured in ambient air, even though some of these compounds (particularly methyl chloride) have much higher atmospheric levels. However, this is not a great drawback because CH_3Cl, CH_2Cl_2, and CH_3Br are the only major halogenated hydrocarbons that require special analytical procedures.

Consequently, electron capture gas chromatography still seems to be the simplest and most generally applicable method for the measurement of environmental levels of halogenated hydrocarbons. It can be used to measure both atmospheric and water concentrations, including samples from municipal water and sewage systems and the marine environment. Furthermore, it allows the measurement of most of these compounds in a single analysis. Ordinary analyses of 5 or 10 ml of ambient air will regularly yield results for fluorocarbons 11, 12 and 113, $CHCl_3$, CH_3CCl_3, CCl_4, $CHCl=CCl_2$, and $CCl_2=CCl_2$ in a single chromatogram, and CH_3Cl, CH_2Cl_2, and fluorocarbon-21 can be added by concentrating 500 or 1000 ml of air before injection into the chromatograph.

Gas chromatography-mass spectrometry (GC-MS) is a more recently developed method for the analysis of halogenated hydrocarbons (Rasmussen 1976; Grimsrud and Rasmussen 1975a, 1975b; Kloepfer and Fairless 1972; Novak et al. 1973; Dowty et al. 1975a, 1975b). One of its advantages over EC-GC is the equal sensitivity GC-MS displays for all compounds. The maximum sensitivity of this method, around 5 parts per trillion v/v (corresponding to 0.5 picograms [0.5×10^{-12} grams] of a compound in a 20-ml sample), represents the sensitivity for all compounds, not just strongly electronegative compounds and halocarbons. It is for this reason that methyl chloride was first identified by this method as a constituent of the atmosphere (Grimsrud and Rasmussen 1975a, 1975b). The ability of GC-MS to positively identify compounds by their characteristic mass spectra is perhaps its greatest advantage over electron capture gas chromatography. Whereas EC-GC usually relies upon the somewhat imprecise method of retention times to identify compounds, GC-MS provides certain identification and usually is not affected by other compounds which may be eluted from the chromatograph at the same time as the compound being measured. GC-MS also can be used to decipher complex compound mixtures, providing at least qualitative identification of the compounds in such mixtures as municipal water and sewage systems and blood plasma.

The disadvantages of GC-MS are the high cost of the instruments ($70,000 or more) and the nonroutine nature of the analysis. The analyses cannot be carried out in the field because the instrument is not portable and relatively highly trained personnel are required to operate it. However, despite statements to the contrary, the method is not limited to the analysis of one compound per sample. In the Washington State University laboratory, three compounds (fluorocarbons 11 and 12, and methyl chloride) per sample have been routinely analyzed by operating the instrument manually; with an associated computer, the number of compounds measured in each analysis could equal the number measured with electron capture gas chromatography.

Long-path infrared spectroscopy is, in theory, a real-time continuous method for measuring ambient concentrations of halocarbons in the atmosphere. However, the sensitivity of this method allows the detection of compounds only with atmospheric concentrations of 10^{-8} (v/v) or better by direct, continuous methods (Hanst 1971, Hanst et al. 1973). This sensitivity is at least 2 orders of magnitude poorer than is needed to measure atmospheric concentrations of halocarbons in any but the most polluted atmospheres and, consequently, the real-time advantages of the method generally have been sacrificed.

In order to get the sensitivity required to measure ambient levels of the halocarbons, discrete samples must be taken by a rather arduous procedure involving cryogenic collection of whole air (taking at least four hours), pumping off nitrogen and oxygen, removing water vapor and CO_2, and then running the remaining sample through a gas chromatograph to separate the compounds for individual analysis in the IR cell. Efforts have been made to make this measurement technique generally applicable (Rasmussen 1976, Hanst et al. 1975); compounds with atmospheric concentrations as low as 10^{-11} by volume (i.e., 10 ppt v/v) can now be measured. This method has been used to measure the fluorocarbons, CCl_4, CH_3CCl_3, and $CCl_2=CCl_2$. However, methyl chloride has not been detected by this means, even though it is known to be present at fairly high levels. This is because the methyl chloride infrared bands are much weaker than those for most other halogenated compounds.

Other disadvantages of long-path infrared spectroscopy are its high cost ($20,000 to $100,000) and the fact that it is not possible to carry out analyses in the field. This method is, however, useful in providing a check on the results obtained by EC-GC and GC-MS since it operates on an absolute basis and should give accurate absolute values if the calibration is done correctly.

The final method that has been used to determine halocarbon concentrations in the atmosphere is infrared solar spectroscopy (Rasmussen 1976, Murcray et al. 1975). This method uses the sun's spectrum passing through the atmosphere at large zenith angles (i.e., when the sun is near the horizon) to obtain the path length necessary to give sufficient absorption to detect ambient halocarbon levels. To date, this technique has been used to detect fluorocarbons and CCl_4 in the stratosphere. Its advantages are that it gives continuous (though not real-time) data on a remote region of the atmosphere where it is difficult to make direct measurements, and that it provides absolute concentration values, within the limits of uncertainty due to variations in the temperature and pressure of gases through which the incoming radiation has passed.

One of the several shortcomings of this method is the fact that it is limited to the stratosphere because of the strong absorption by water vapor in the troposphere. It also is limited by the strength of the IR absorption bands for some compounds (for example, CH_3Cl probably cannot be measured by this method). Furthermore, the method is not routine, has little flexibility, takes considerable time to transform collected data into quantitative numbers, and requires computer analysis. Therefore, while it is an excellent method to obtain quantitative values for the concentration of selected halocarbons in the stratosphere, it is limited to that function alone.

Water

Many of the techniques and problems in analyzing halogenated methanes in water are similar to those in air analyses. Gas chromatography with electron capture (EC-GC) or other detectors is most widely used, and absolute or confirmatory analysis is performed with mass spectrometry (GC-MS)(Keith 1976). Although direct aqueous injection occasionally is used in such gas chromatography analyses (McKinney et al. 1976, Nicholson and Meresy 1975), it usually is necessary to concentrate the sample before analysis. This has been done in various ways, including carbon adsorption, solvent extraction, freeze drying, steam distillation, reverse osmosis, inert gas stripping, and resin sorption (Keith 1976).

In one of the earliest studies of halogenated methanes in a municipal water system, Rook (1974) used a static headspace analysis, and analyzed samples of the equilibrated atmosphere over a solution in a closed container. Bellar and Lichtenberg (1974) concentrated the aqueous sample by continuous purging with an inert gas, which transferred the volatile organics to a porous polymer trap. Then the compounds were desorbed by heating at 125-130°C and transfered to a GC for analysis. This technique was found to have a lower limit of detection than the static headspace analysis for halogenated methanes, in the range of 1.0 µg/l (Bellar and Lichtenberg 1974).

This method has been used with some modification in the EPA National Organics Reconnaissance Survey (NORS) of halogenated methanes in municipal water supplies (Kopfler et al. 1976). Henderson et al. (1976) have noted, however, that there are some disadvantages to the purge-adsorption technique, such as the need for special equipment, the time required for analysis, and reported difficulties that are due in part to thermal and/or hydrolytic degradation on the adsorption column. They described a simple liquid-liquid extraction technique in which the extracting organic solvent was directly injected through the septum of a serum bottle with an aqueous sample. After extraction the concentrate was similarly withdrawn and was analyzed by EC-GC. Using 1,2-dibromoethane as an internal standard, Henderson et al. demonstrated the utility of the technique in analyzing raw and treated water from a municipal supply.

Although direct passage of an aqueous solution onto a resin column may be useful for concentrating nonvolatile organics, a large amount of the volatiles, such as chloroform, is likely to be lost in the process (Glaze et al. 1976). However, such difficulties are not unique to direct resin column concentration procedures; great care must be taken in all techniques to avoid volatilization losses.

Most existing studies of halogenated methanes employ techniques that of necessity involve single sample analyses. For some purposes, it may be useful to continuously monitor these compounds. Attempts of this kind have been reported by Mieure et al. (1976) in which membrane techniques were used in conjunction with mass spectrometry. In one technique, the investigators continuously passed an aqueous solution across a silicone, semipermeable membrane interfaced with a mass spectrometer. Up to four compounds could be monitored simultaneously and the detection limits for chloroform and methylene chloride each were 4.0 µg/l. A second technique used a dialysis membrane by passing the aqueous solution continuously across it, the organics then dialyzing to a solvent on the other side. The time-averaged concentrate was then analyzed by EC-GC. This technique was used to analyze chloroform, carbon tetrachloride, dichlorobromomethane, and chlorodibromomethane in municipal tap water.

Within the past few years, much interest has developed in the significance and source of halogenated methanes largely because of the analytical chemist's ability to reliably detect and quantify them. Although the procedures cannot yet be considered routine, many of the difficulties have been recognized and overcome to the point where many laboratories are analyzing halogenated methanes in water with confidence.

Given the strengths and weaknesses of the four primary methods for measuring ambient concentrations of halogenated hydrocarbons in the environment, electron capture-gas chromatography and gas chromatography-mass spectrometry are currently the most practical, versatile, and economical methods available. They can be used for the analysis of both atmospheric and water samples; they are extremely sensitive in routine analyses; and when used in conjunction with one another, they are complementary. Infrared spectroscopy should be pursued further in an attempt to make it more generally applicable, because it is inherently more accurate than the other methods. There also are such possibilities on the horizon as the atmospheric pressure ionization mass spectrometer, which could greatly increase the power of GC-MS. Other methods for measuring halocarbons are still in the physical chemistry state-of-the-art stage, and are not of practical importance at this time.

Sampling and Contamination

All methods used to date to analyze samples of air and water for their halogenated hydrocarbon content have experienced problems of contamination, adsorption, and degradation. Constant attention to the elimination of these problems is one of the major efforts required of any halocarbon measurement program.

There are four general approaches used to collect discrete samples of air for analysis of trace gas concentrations. These are:

1. Cryogenic sampling, in which liquid helium or liquid nitrogen is used to cool a container to extremely low temperatures. Air is pumped into the container, where it liquifies and creates a partial vacuum that the pump continues to fill. This is continued until the maximum safe amount of air has been pumped into the container. The method permits the collection of enormous amounts of air, often several thousand liters.

2. Pump-pressurized samples, in which a mechanical pump is used without cryogenic assistance to fill a container to a positive pressure relative to the surrounding atmosphere. The pressure that can be attained by this method depends upon the particular pump used; usually, the sample volume collected is far less than with cryogenic assistance. However, for analytical methods in which sensitivity is not a problem (such as electron capture-gas chromatography), this procedure can provide an ample amount of air for analysis and is much simpler and quicker than cryogenic sampling.

3. Ambient or subambient pressure sampling, in which an evacuated container is simply opened and allowed to fill until it has reached ambient pressure at the sampling location. Since many samples are taken at high altitudes, containers are often at below-ambient pressure when they are returned to the ground for analysis. Although this method is the simplest and most straightforward sampling procedure available, subambient pressure samples are prone to contamination through miniscule leaks in containers as well as through the connecting transfer lines. Also, the subambient pressure of the samples limits the amount of material available for analysis.

4. Adsorption of selected gases on such adsorbents as molecular sieves, activated charcoal, and the like also is used to collect samples. This method permits selective sampling of large volumes of air, but it also has many drawbacks, especially in desorption and quantitative assays.

Obviously, the ideal way to measure a component of the atmosphere is _in situ_, thereby obviating the need for arduous sampling procedures and eliminating many of the problems of contamination, degradation, and adsorption. Unfortunately, this kind of analysis is often impossible, particularly when used with such methods as GC-MS and GC-IR, which require large, complicated, sensitive equipment that does not readily lend itself to field operation. The ability to make direct, continuous (or quasi-continuous) measurements is one of the chief attractions of long-path infrared spectroscopy and is also an added bonus of electron capture gas chromatography. Highly specialized miniature mass spectrometers and electron capture-gas chromatographs are being designed to operate during high-altitude balloon flights to meet an urgent need for _in situ_ data on the stratospheric concentration distribution of the halocarbons.

Of the four general methods used to collect discrete samples, it has been found that contamination and other problems are more serious in low-pressure sampling than in high-pressure sampling systems. The reason, is that in a low-pressure sample there is less sample, so that it takes less to contaminate it, and the adsorption of less volatile compounds on the vessel walls is more significant, because the surface-to-mass ratio is so much larger than in high-pressure systems. In a sample pressurized to 100 psi or more, there is such a large mass of sample present that an equivalent amount of contamination or adsorption would be insignificant.

Concerns about sampling problems were discussed at length at the Halocarbon Workshop sponsored in 1976 (Rasmussen 1976) by the National Aeronautics and Space Administration, the National Bureau of Standards, and the National Science Foundation. In general, the conclusion of the workshop panel considering contamination was: "Can you really defend the position that your sample is not contaminated?" Defense of the integrity of a sample requires verification that every part of the sampling system is free from contamination and that a sample of air can be stored successfully for a time equal to or greater than the time that it would take to analyze real samples. The entire system, including the sample container, should be checked using either a sample of ambient air of known composition or a synthetic mixture that is as close as possible to the ambient air in its composition, concentration, and pressure.

Water samples are subject to the same possibility of contamination and other problems that exist in air sampling. Particular care must be devoted to the sampling, transportation, and storage of aqueous samples of halogenated methanes because of their volatility and the complexity of the samples, especially those containing chlorine or other oxidants. A technique that is often used involves filling and sealing a serum bottle without air space and storing it just above freezing (Kopfler et al. 1976), because at higher storage temperatures, the chloroform and bromodichloromethane concentrations increase with time in chlorinated water samples. This problem has also been partially eliminated by reducing the chlorine with potassium ferrocyanide. Sodium thiosulfate, commonly used to reduce chlorine, was not used because of its ability to react with alkyl halides.

Water samples generally require additional preparation before analysis. Normally, they may be concentrated by various water analysis techniques but direct aqueous injection is used occasionally in GC analyses.

Calibration

At present, there are no primary calibration standards for the nonfluorinated halomethanes and it is unlikely that any accepted primary standard will be available in the next few years because no agency has taken responsibility for producing one that is absolute. However, it is expected that the National Bureau of Standards will develop several primary fluorocarbon standards within the next two years.

As for secondary standards, several groups already have taken steps to prepare such standards to give their measurements long-term internal consistency. These standards are not necessarily absolutely accurate, but they will allow groups to reference future measurements against those of the past and present, and, ultimately, against an absolute standard when one is developed.

Attempts at inter-laboratory calibration have been made by Washington State University. At the 1976 Halocarbon Workshop (Rasmussen 1976), the results indicated that for some compounds, notably the fluorocarbons, the agreement among most of the laboratories was within ±10 percent. Most compounds under consideration in this report were not among those measured. The exception, carbon tetrachloride (CCl_4), was unfortunately the compound which displayed the greatest variation among participating laboratories. Concentration values for CCl_4 in the same sample of air varied by as much as a factor of 4 among the different laboratories. Obviously, there is still much work to be done before absolute values can be assigned to ambient concentrations of many of the halogenated hydrocarbons. At present, all that is known with certainty is that we are within an order of magnitude of the absolute value of these compounds.

NOTES

1. Davis, D.D. (1975) unpublished results, cited in Hampson, R.F. and D. Garvin (1975) National Bureau of Standards Technical Note 866. Washington, D.C.: U.S. Department of Commerce.

2. Davis, D.D. et al. (1975) private communication cited in Yung et al. (1975).

3. Howard, C.J. and K.M. Evenson (1975) private communication cited in Yung et al. (1975).

4. Howard, C.J. (1975) unpublished results, cited in Hampson, R.F. and D. Garvin (1975) National Bureau of Standards Technical Note 866. Washington, D.C.: U.S. Department of Commerce.

REFERENCES

Atkinson, R., G.M. Breuer, J.N. Pitts, Jr., and H.L. Sandoval (1976) Tropospheric and stratospheric sinks for halocarbons: Photooxidation, O('D) atom, and OH radical reactions. Journal of Geophysical Research 81(33):5765-5770.

Austin, G.T. (1974) Industrially significant organic chemicals, Part 3. Chemical Engineering 81(6):87-92.

Bellar, T.A. and J.J. Lichtenberg (1974) Determining volatile organics at microgram-per-liter levels by gas chromatography. Journal of the American Water Works Association 66(12):739-744.

Cox, R.A., R.G. Derwent, A.E.J. Eggleton, and J.E. Lovelock (1976) Photochemical oxidation of halocarbons in the troposphere. Atmospheric Environment 10(4):305-308.

Dilling, W.L., N.B. Tefertiller, and G.J. Kallos (1975) Evaporation rates and reactivities of methylene chloride, chloroform, 1,1,1-trichloroethane, trichloroethylene, tetrachloroethylene, and other chlorinated compounds in dilute aqueous solutions. Environmental Science and Technology 9:833-838.

Dowty, B., D. Carlisle, J.L. Laseter, and J. Storer (1975a) Halogenated hydrocarbons in New Orleans drinking water and blood plasma. Science 187:75-77.

Dowty, B.J., D.R. Carlisle, and J.L. Laseter (1975b) New Orleans drinking water sources tested by gas chromatography-mass spectrometry. Environmental Science and Technology 9(8):762-765.

Faith, W.L., O.B. Keyes, and R.L. Clark (1965) Industrial Chemicals, 3rd ed. New York: John Wiley and Sons, Inc.

Glaze, W.H., J.E. Henderson, IV, and G. Smith (1976) Pages 247-254, Identification and Analysis of Organic Pollutants in Water, edited by L. Keith. Ann Arbor, Mich.: Ann Arbor Science Publishers, Inc.

Goldman, P. (1972) Enzymology of carbon-halogen bonds. In Degradation of Synthetic Organic Molecules in the Biosphere. Washington, D.C.: National Academy of Sciences.

Gordon, J. (1976) Air Pollution Assessment of Methylene Chloride, Mitre Technical Report MTR-7334. McLean, Va.: Mitre Corp.

Grimsrud, E.P. and R.A. Rasmussen (1975a) The analysis of chlorofluorocarbons in the troposphere by gas chromatography-mass spectrometry. Atmospheric Environment 9:1010-1013.

Grimsrud, E.P. and R.A. Rasmussen (1975b) Survey and Analysis of halocarbons in the atmosphere by gas chromatography-mass spectrometry. Atmospheric Environment 9:1014-1017.

Hanst, P.L. (1971) Spectroscopic measurements for air pollution measurements. Pages 91-213, Advances in Environmental Science and Technology, Vol. 2, edited by J.N. Pitts and R.L. Metcalf. New York: John Wiley and Sons, Inc.

Hanst, P.L., A.S. Lefohn, and B.W. Gay, Jr. (1973) Detection of atmospheric pollutants at parts-per-billion levels by infrared spectroscopy. Applied Spectroscopy 27:188.

Hanst, P.L., L.L. Spiller, D.M. Watts, J.W. Spence, and M.F. Miller (1975) Infrared measurement of fluorocarbons, carbon tetrachloride, carbonyl sulfide, and other atmospheric trace gases. Journal of the Air Pollution Control Association 25(12):1220-1226.

Henderson, J.E., G.R. Peyton, and W.H. Glaze (1976) Pages 105-112, Identification and Analysis of Organic Pollutants in Water, edited by L. Keith. Ann Arbor, Mich.: Ann Arbor Science Publishers, Inc.

Heukelakian, H.H. and M.C. Rand (1955) Biochemical oxygen demand of pure organic compounds. Sewage and Industrial Wastes 27:1040-1053.

Johns, R. (1976) Air Pollution Assessment of Carbon Tetrachloride, Mitre Technical Report No. MTR-7144. McLean, Va.: Mitre Corp.

Keith, L., ed. (1976) Identification and Analysis of Organic Pollutants in Water. Ann Arbor, Mich.: Ann Arbor Science Publishers, Inc.

Kloepfer, R.D. and B.J. Fairless (1972) Characterization of organic components in a municipal water supply. Environmental Science and Technology 6(12):1036-1037.

Kopfler, F.C., R.G. Melton, R.D. Lingg, and W.E. Coleman (1976) Pages 87-104, Identification and Analysis of Organic Pollutants in Water, edited by L. Keith. Ann Arbor, Mich.: Ann Arbor Science Publishers, Inc.

Krey, P.W., R. Lagomarsino, M. Schonberg, and I.E. Toonkel (1976) Trace gases in the stratosphere. Pages 21-49, Health and Safety Laboratory Environmental Quarterly April 1, U.S. Energy Research and Development Agency, prepared by E.P. Hardy, Jr. Springfield, Va.: National Technical Information Service.

Lillian, D., H.B. Singh, A. Appleby, L. Lobban, R. Arnts, R. Gumpert, R. Hague, J. Toomey, J. Kazazis, M. Antell, D. Hansen, and B. Scott (1975) Atmospheric fates of halogenated compounds. Environmental Science and Technology 9(12):1042-1048.

Lovelock, J.E. (1975) Natural halocarbons in the air and in the sea. Nature 256:193-194.

Lovelock, J.E., R.J. Maggs, and R.J. Wade (1973) Halogenated hydrocarbons in and over the Atlantic. Nature 241:194-196.

Mabey, W.R. and T. Mill (1976) Kinetics of Hydrolysis and Oxidation of Organic Pollutants in the Aquatic Environment. Extended abstracts of Symposium on Nonbiological Transport and Transformation of Pollutants on Land and Water: Processes and Critical Data Required for Predictive Description, May 11-13, 1976. Gaithersburg, Md.: National Bureau of Standards.

McConnell, G., D.M. Ferguson, and C.R. Pearson (1975) Chlorinated hydrocarbons and the environment. Endeavour 34:13-18.

McKinney, J.D., R.R. Maurer, J.R. Hass, and R.O. Thomas (1976) Pages 417-432, Identification and Analysis of Organic Pollutants in Water, edited by L. Keith. Ann Arbor, Mich.: Ann Arbor Science Publishers, Inc.

Mieure, J.P., G.W. Mappes, E.S. Tucker, and M.W. Dietrich (1976) Pages 113-133, Identification and Analysis of Organic Pollutants in Water, edited by L. Keith. Ann Arbor, Mich.: Ann Arbor Science Publishers, Inc.

Moelwyn-Hughes, E.A. (1957) Physical Chemistry. New York: Pergamon Press.

Molina, M.J. and F.S. Rowland (1974) Predicted present stratospheric abundances of chlorine species from photodissociation of carbon tetrachloride. Geophysical Research Letters 1(7):309-312.

Montgomery, H.A.C. and M. Conlon (1967) Detection of chlorinated solvents in sewage sludge. Water Pollution Control 66(2):190-192.

Murcray, D.G., F.S. Bonomo, J.N. Brooks, A. Goldman, F.H. Murcray, and W.J. Williams (1975) Detection of fluorocarbons in the stratosphere. Geophysical Research Letters 2(3):109-112.

Murray, A.J. and J.P. Riley (1973a) Determination of chlorinated aliphatic hydrocarbons in air, natural waters, marine organisms, and sediments. Analytica Chimica Acta 65:261-270.

Murray, A.J. and J.P. Riley (1973b) Occurrence of some chlorinated aliphatic hydrocarbons in the environment. Nature 242:37-38.

Neely, W.B., G.E. Blau, and T. Alfrey, Jr. (1976) Mathematical models predict concentration-time profiles from chemical spill in a river. Environmental Science and Technology 10:72-76.

Nicholson, A.A. and O. Meresy (1975) Analysis of volatile, halogenated organics in water by direct aqueous injection-gas chromatography. Bulletin of Environmental Contamination and Toxicology 14(4):453-455.

Novak, J., J. Zluticky, V. Kubelka, and J. Mostecky (1973) Analysis of organic constituents present in drinking water. Journal of Chromatography 76(1):45-50.

Pearson, C.R. and G. McConnell (1975) Proceedings of the Royal Society of London B 189:305-332.

Rasmussen, R.A., ed. (1976) Report of the Workshop on Halocarbon Analysis and Measurement. Halocarbon Workshop sponsored by NASA-NBS-NSF, Boulder Colorado, March 25-26.

Robbins, D.E. (1976) Photodissociation of methyl chloride and methyl bromide in the atmosphere. Geophysical Research Letters 3(4):213-216.

Rook, J.J. (1974) Journal of the Society of Water Treatment and Examination 23 (Part 2):234-243.

Singh, H.B. (1976) Distribution, sources and sinks of atmospheric halogenated compounds. Paper for presentation at the 69th Annual Meeting of the Air Pollution Control Association, June 27-July 1, Portland, Oregon. (Abstract)

Singh, H.B., L. Salas, A. Crawford, P.L. Hanst, and J.W. Spence (1977) Urban-non-urban relationships of halocarbons, SF_6, N_2O and other atmospheric contitutents. Atmospheric Environment 11(9):819.

Spence, J.W., P.L. Hanst, and B.W. Gay, Jr. (1976) Atmospheric oxidation of methyl chloride, methylene chloride, and chloroform. Journal of the Air Pollution Control Association 26(10):994-996.

Standen, A., ed. (1964) Kirk-Othmer Encyclopedia of Chemical Technology, 2nd ed. New York: Interscience Publishers.

Su, C. and E.D. Goldberg (1976) Environmental concentrations and fluxes of some halocarbons. Chapter 14, Marine Pollutant Transfer, edited by H.L. Windom and R.A. Duce. Lexington, Mass.: Lexington Books, D.C. Heath and Company.

Williams, I.H. (1965) Gas chromatographic techniques for the identification of low concentrations of atmospheric pollutants. Analytical Chemistry 37(13):1723-1732.

Wolfe, N.L., R.G. Zepp, G.L. Baughman, R.C. Fincher, and J.A. Gordon (1976) Chemical and Photochemical Transformations of Selected Pesticides in Aquatic Systems. EPA-600/3-76-067. Washington, D.C.: U.S. Environmental Protection Agency.

Yaws, C.L. (1976) Methyl chloride, methylene chloride, chloroform and carbon tetrachloride. Chemical Engineering 83(14):81-89.

Yung, Y.L., M.B. McElroy, and S.C. Wofsy (1975) Atmospheric halocarbons: a discussion with emphasis on chloroform. Geophysical Research Letters 2(9):397-399.

CHAPTER 4

PRODUCTION AND USES

PRODUCTION OF HALOMETHANES

The chloromethanes represent the most significant commercial volumes of the halomethanes, except for those halomethanes containing fluorine. McCarthy (1974) has estimated that of the total chlorine produced in the United States from 1963 to 1972, 115 million metric tons, about 35 percent was used in the production of chlorinated organic compounds. The chloromethanes produced in this period contained 6 million metric tons or about 5 percent of the total chlorine consumed for all uses. Other chlorine-containing organics of significant volume included: polyvinyl chloride (the major chlorine consumer), perchloroethylene, trichloroethylene, 1,1,1-trichloroethane, ethyl chloride, ethylene dichloride, chlorofluorocarbons, chlorinated paraffins, and chlorinated aromatics.

U.S. Production of Chloromethanes

Production of the four principal chloromethanes--carbon tetrachloride, chloroform, methylene chloride, and methyl chloride--is interrelated because of co-production facilities and processes. Direct chlorination of methane can produce all four compounds, with some variations in the proportions of each one depending upon reaction conditions selected to meet market demands. Hydrochlorination of methanol to methyl chloride also can be integrated with subsequent chlorination. In addition, carbon tetrachloride is produced independently, along with perchloroethylene, by chlorination of propylene.

The U.S. production histories of chloromethanes are compiled by the U.S. International Trade Commission (1976) and tabulated summaries and analyses of production and sales of chloromethanes are published annually by the Stanford Research Institute (1976a). Actual U.S. production of the chloromethanes began in 1908 by a sole producer of carbon tetrachloride but reports on production did not begin until 1922, when there were multiple producers of this compound.

Annual U.S. production of the four chloromethanes and the percentage of annual totals for each species are summarized in Table 4.1. Figure 4.1 depicts the annual production graphically. Before 1940, carbon tetrachloride was the predominant product, accounting for more than 90 percent of the annual chloromethane market. This share fell to less than 60 percent by 1960 as new uses were developed for the other three compounds. Currently, carbon tetrachloride represents 41 percent of the total annual market volume; chloroform, 14 percent; methylene chloride, 26 percent; and methyl chloride, 18 percent.

Carbon tetrachloride production grew at an annual rate of about 15 percent in the United States from 1922 to 1945, when the compound was used extensively as an industrial solvent, fire extinguishing fluid, and dry-cleaning agent. Recognition of its toxicity led to its gradual replacement with less toxic chlorinated ethanes and ethylenes, and other chloromethanes. In most of the postwar years from 1945 to 1960, carbon tetrachloride production grew at only 5 percent per year, but increased to about 9.5 percent from 1960 to 1970 as a result of the growing fluorocarbon market. Growth in the United States leveled off after 1970 as the remaining dispersive uses were largely replaced by other solvents and as uncertainties appeared in the fluorocarbon aerosol market because of the controversy over ozone depletion.

Production growth rates of the other three chloromethanes in the United States were roughly parallel to each other from 1940 to 1965. Annual growth of chloroform in the United States was 15 percent during this period but has fallen off to 5 percent since 1965. Chloroform is predominantly a chemical intermediate for fluorocarbons used as refrigerants rather than as aerosol propellants. However, chloroform is under scrutiny for toxicological reasons and has been replaced in anesthetics, cough syrups, and other pharmaceutical formulations.

Methylene chloride production has experienced a growth of about 11 percent a year since 1950. It has the most diverse uses of any chloromethane and recently has replaced trichloroethylene in the extraction of caffeine from coffee. Annual U.S. growth rates of methyl chloride production have fluctuated more than those of any other chloromethane. It is used predominantly as a chemical intermediate.

Cumulative U.S. production of chloromethanes is shown in Table 4.2. Figure 4.2 is a graphic presentation of the cumulative production data. Cumulative production of carbon tetrachloride has been extrapolated from 1908. The U.S. cumulative production for all chloromethanes through 1976 is estimated to be 15×10^6 metric tons of which 54 percent is

TABLE 4.1 Annual U.S. Production of Chloromethanes (10^3 metric tons) and Percentage for Each Species, 1908-1976

Year	Carbon Tetrachloride (CCl_4)		Chloroform ($CHCl_3$)		Methylene Chloride (CH_2Cl_2)		Methyl Chloride (CH_3Cl)		Total U.S. Annual Production
	Annual Production	Percent of Annual Total	Annual Production	Percent of Annual Total	Annual Production	Percent of Annual Total	Annual Production	Percent of Annual Total	
1908	0.09E	100.0	—	—	—	—	—	—	0.09
1909	0.27E	100.0	—	—	—	—	—	—	0.27
1910	0.23E	100.0	—	—	—	—	—	—	0.23
1911	0.68E	100.0	—	—	—	—	—	—	0.68
1912	0.77E	100.0	—	—	—	—	—	—	0.77
1913	1.0 E	100.0	—	—	—	—	—	—	1.0
1914	1.5 E	100.0	—	—	—	—	—	—	1.5
1915	2.8 E	100.0	—	—	—	—	—	—	2.8
1916	5.1 E	100.0	—	—	—	—	—	—	5.1
1917	5.9 E	100.0	—	—	—	—	—	—	5.9
1918	6.4 E	100.0	—	—	—	—	—	—	6.4
1919	3.3 E	100.0	—	—	—	—	—	—	3.3
1920	3.2 E	100.0	—	—	—	—	—	—	3.2
1921	1.9 E	100.0	—	—	—	—	—	—	1.9
1922	5.1	100.0	—	—	—	—	—	—	5.1
1923	6.1	89.4	0.73	10.6	—	—	—	—	6.8
1924	6.5	91.7	0.59	8.33	—	—	—	—	7.1
1925	7.3	92.6	0.59	7.42	—	—	—	—	7.9
1926	8.6	90.9	0.86	9.08	—	—	—	—	9.5
1927	7.5	88.3	1.0 E	11.7	—	—	—	—	8.5
1928	8.9	89.1	1.1 E	10.9	—	—	—	—	10
1929	16	92.5	1.3	7.47	—	—	—	—	17
1930	16	93.2	1.1	6.79	—	—	—	—	17
1931	16E	93.8	1.0 E	6.20	—	—	—	—	17
1932	16E	94.2	1.0 E	5.83	—	—	—	—	17
1933	16	94.4	0.95	5.55	—	—	—	—	17
1934	21	96.8	0.82	3.69	—	—	—	—	22
1935	25	93.5	0.86	3.18	—	—	0.91	3.35	27
1936	31	92.3	1.0 E	2.98	—	—	1.4	4.20	33
1937	38	93.2	1.2	2.99	—	—	1.5	3.77	41
1938	35	93.6	1.0	2.64	—	—	1.4	3.72	38
1939	41	93.9	1.3	3.01	—	—	1.4	3.11	44

58

Year								
1940	46	91.6	1.4	2.82	1.5E	2.91	1.4	2.72	50
1941	55	88.8	2.9	4.74	1.8E	2.91	2.2	3.57	62
1942	66	89.4	3.4E	4.63	2.3E	3.15	2.1	2.84	73
1943	80	86.8	3.9	4.31	2.9E	3.22	5.2	5.69	92
1944	95	83.2	4.4	3.89	3.8	3.29	11	9.64	114
1945	87	79.5	4.2	3.79	4.9E	4.45	13	12.2	110
1946	67	73.7	4.9	5.30	6.4	6.93	13	14.1	92
1947	90	77.5	5.5	4.74	8.5	7.31	12	10.5	117
1948	97	79.8	5.7	4.68	8.2	6.72	11E	8.84	122
1949	89	74.4	6.2	5.17	15	12.6	9.4	7.85	120
1950	98	71.3	9.2	6.68	18	13.1	12	8.92	138
1951	111	70.4	12	7.49	18	11.5	17	10.6	157
1952	100	66.5	10	6.67	25	16.7	15	10.2	150
1953	118	66.7	12	6.55	29	16.3	18	10.4	177
1954	107	63.5	15	8.68	32	18.9	15	8.95	168
1955	130	65.6	18	9.22	34	16.9	16	8.29	199
1956	137	62.3	21	9.53	43	19.6	19	8.56	220
1957	144	61.6	26	11.1	43	18.2	21	9.07	234
1958	142	63.5	21	9.56	40	18.1	20	8.83	223
1959	167	59.5	32	11.4	51	18.2	30	10.9	280
1960	169	57.6	35	11.8	51	17.5	38	13.0	293
1961	174	56.3	35	11.3	53	17.0	48	15.4	309
1962	219	58.0	45	11.8	65	17.3	49	12.9	378
1963	236	58.6	48	11.9	67	16.7	52	12.9	402
1964	243	55.3	54	12.3	81	18.5	61	13.8	439
1965	269	51.9	69	13.3	96	18.4	85	16.4	519
1966	294	48.7	81	13.4	121	20.1	107	17.8	604
1967	324	49.5	87	13.2	119	18.2	125	19.1	654
1968	346	49.2	82	11.6	137	19.5	138	19.7	704
1969	400	47.3	98	11.6	166	19.6	183	21.6	847
1970	459	48.7	109	11.6	182	19.4	192	20.4	942
1971	458	48.5	105	11.1	182	19.3	198	21.0	943
1972	452	46.2	106	10.9	214	21.9	206	21.0	978
1973	475	44.3	115	10.7	236	22.0	247	23.0	1072
1974	527P	45.3	137P	11.8	276P	23.7	224P	19.2	1164
1975	413P	46.3	120P	13.4	221P	24.8	139P	15.5	893
1976	388P	41.4	133P	14.2	247P	26.4	167P	17.9	935

NOTE: E—Panel estimates; P—Preliminary total.
SOURCE: U.S. International Trade Commission (1976).

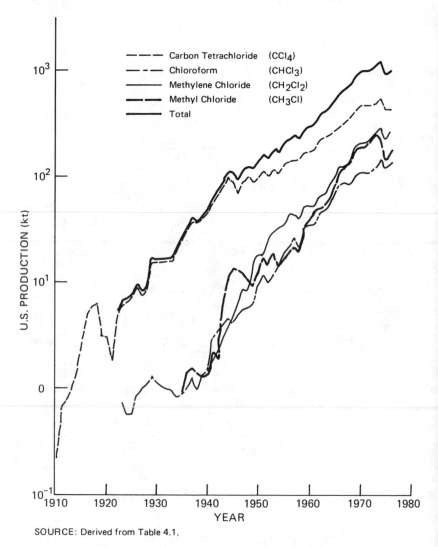

FIGURE 4.1 Annual U.S. production of chloromethanes (10^3 metric tons), 1908-1976.

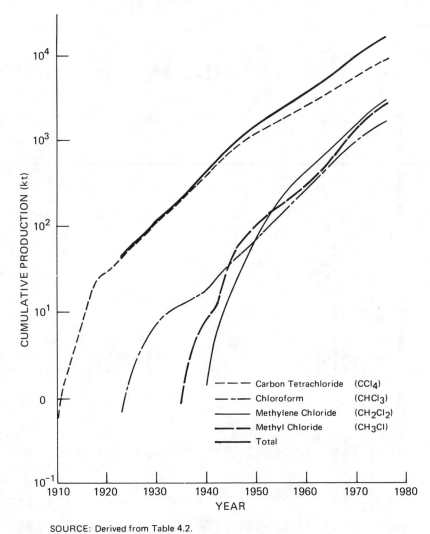

SOURCE: Derived from Table 4.2.

FIGURE 4.2 Cumulative U.S. production of chloromethanes (10^3 metric tons), 1908-1976.

TABLE 4.2 Cumulative U.S. Production of Chloromethanes (10^3 metric tons) and Cumulative Percentage of Each Species, 1908-1976

Year	Carbon Tetrachloride (CCl$_4$) Cumulative Production Total	Carbon Tetrachloride (CCl$_4$) Percent of Cumulative Total	Chloroform (CHCl$_3$) Cumulative Production Total	Chloroform (CHCl$_3$) Percent of Cumulative Total	Methylene Chloride (CH$_2$Cl$_2$) Cumulative Production Total	Methylene Chloride (CH$_2$Cl$_2$) Percent of Cumulative Total	Methyl Chloride (CH$_3$Cl) Cumulative Production Total	Methyl Chloride (CH$_3$Cl) Percent of Cumulative Total	Cumulative Total U.S. Production	Percent of Cumulative U.S. Production
1908	0.09	0.001	—	—	—	—	—	—	0.09	0.001
1909	0.36	0.002	—	—	—	—	—	—	0.36	0.002
1910	0.59	0.001	—	—	—	—	—	—	0.59	0.001
1911	1.3	0.004	—	—	—	—	—	—	1.3	0.004
1912	2.0	0.005	—	—	—	—	—	—	2.0	0.005
1913	3.1	0.007	—	—	—	—	—	—	3.1	0.007
1914	4.5	0.009	—	—	—	—	—	—	4.5	0.009
1915	7.3	0.018	—	—	—	—	—	—	7.3	0.018
1916	12	0.033	—	—	—	—	—	—	12	0.033
1917	18	0.039	—	—	—	—	—	—	18	0.039
1918	25	0.042	—	—	—	—	—	—	25	0.042
1919	28	0.021	—	—	—	—	—	—	28	0.021
1920	31	0.021	—	—	—	—	—	—	31	0.021
1921	33	0.012	—	—	—	—	—	—	33	0.012
1922	38	0.033	—	—	—	—	—	—	38	0.033
1923	44	0.040	0.73	0.005	—	—	—	—	45	0.045
1924	51	0.042	1.3	0.004	—	—	—	—	52	0.046
1925	58	0.048	1.9	0.004	—	—	—	—	60	0.052
1926	67	0.056	2.8	0.006	—	—	—	—	69	0.062
1927	74	0.049	3.8	0.006	—	—	—	—	78	0.055
1928	83	0.058	4.8	0.007	—	—	—	—	88	0.065
1929	99	0.102	6.1	0.008	—	—	—	—	105	0.111
1930	114	0.101	7.3	0.007	—	—	—	—	122	0.109
1931	130	0.103	8.3	0.007	—	—	—	—	139	0.109
1932	146	0.105	9.3	0.006	—	—	—	—	156	0.111
1933	163	0.105	10	0.006	—	—	—	—	173	0.112
1934	184	0.138	11	0.005	—	—	—	—	195	0.144
1935	209	0.165	12	0.006	—	—	0.91	0.006	222	0.176
1936	240	0.202	13	0.006	—	—	2.3	0.009	255	0.218
1937	278	0.248	14	0.008	—	—	3.9	0.010	296	0.266
1938	314	0.230	15	0.006	—	—	5.3	0.009	334	0.246
1939	355	0.267	16	0.009	—	—	6.6	0.009	378	0.285

62

Year										
1940	400	0.298	18	0.009	1.5	0.009	8.0	0.009	428	0.325
1941	456	0.360	21	0.019	3.3	0.012	10	0.014	490	0.405
1942	521	0.427	24	0.022	5.6	0.015	12	0.014	563	0.478
1943	601	0.517	28	0.026	8.5	0.019	18	0.034	655	0.596
1944	696	0.619	33	0.029	12	0.024	29	0.072	769	0.744
1945	783	0.569	37	0.027	17	0.032	42	0.088	879	0.716
1946	851	0.439	42	0.032	24	0.041	55	0.084	971	0.596
1947	941	0.588	47	0.036	32	0.055	67	0.079	1088	0.759
1948	1039	0.634	53	0.037	40	0.053	78	0.070	1210	0.795
1949	1128	0.581	59	0.040	55	0.099	87	0.061	1330	0.782
1950	1226	0.639	68	0.060	73	0.117	100	0.080	1468	0.896
1951	1337	0.722	80	0.077	92	0.118	116	0.109	1625	1.02
1952	1437	0.648	90	0.065	116	0.162	132	0.099	1775	0.974
1953	1555	0.767	102	0.075	145	0.188	150	0.120	1951	1.15
1954	1661	0.693	116	0.095	177	0.206	165	0.098	2119	1.09
1955	1791	0.848	135	0.119	211	0.218	181	0.107	2318	1.29
1956	1929	0.893	156	0.137	254	0.282	200	0.123	2538	1.43
1957	2073	0.938	182	0.169	297	0.278	222	0.138	2772	1.52
1958	2215	0.924	203	0.139	337	0.263	241	0.128	2996	1.45
1959	2382	1.09	235	0.209	388	0.333	272	0.198	3276	1.82
1960	2550	1.10	270	0.225	439	0.334	310	0.249	3569	1.91
1961	2724	1.13	305	0.228	492	0.342	358	0.310	3879	2.01
1962	2944	1.43	349	0.290	557	0.424	406	0.318	4257	2.46
1963	3179	1.53	397	0.311	624	0.437	458	0.336	4659	2.62
1964	3422	1.58	451	0.352	706	0.530	519	0.396	5098	2.86
1965	3692	1.75	520	0.450	801	0.622	604	0.553	5617	3.38
1966	3986	1.91	601	0.528	923	0.789	711	0.699	6221	3.93
1967	4309	2.11	688	0.564	1041	0.774	836	0.813	6875	4.26
1968	4656	2.25	770	0.534	1179	0.893	975	0.901	7579	4.58
1969	5056	2.61	868	0.638	1345	1.08	1158	1.19	8426	5.51
1970	5515	2.98	977	0.708	1527	1.19	1349	1.25	9368	6.13
1971	5972	2.98	1082	0.681	1709	1.18	1548	1.29	10311	6.14
1972	6424	2.94	1188	0.693	1923	1.39	1753	1.34	11289	6.36
1973	6899	3.09	1303	0.746	2159	1.54	2000	1.61	12361	6.98
1974	7427	3.43	1440	0.891	2435	1.80	2224	1.46	13525	7.57
1975	7840	2.69	1559	0.779	2656	1.44	2363	0.903	14418	5.81
1976	8228	2.53	1692	0.866	2903	1.61	2530	1.09	15353	6.09
Total Percent		53.6		11.0		18.9		16.15		100.0

SOURCE: U.S. International Trade Commission (1976).

carbon tetrachloride, 11 percent is chloroform, 19 percent is methylene chloride, and 16 percent is methyl chloride.

World Production of Chloromethanes

The production estimates employed by several authors (Altshuller 1976, Singh et al. 1976, Galbally 1976) and those used in this study are essentially the same. Panel estimates of "free world" chloromethane production for 1976 are summarized in Table 4.3. The United States accounted for about 54 percent of "free world" chloromethane production in 1976. This figure agrees well with Galbally's estimate that in 1973 U.S. production was about 50 percent of annual "free world" production (Galbally 1976). Singh et al. (1976) estimated the total cumulative production of CCl_4 in Japan and Western Europe to be about 100 metric tons before 1960. Since then, CCl_4 production has grown rapidly in both Japan and Western Europe until their combined production accounted for about 45 percent of "free world" production in 1976.

Information on the production and uses of halomethanes is not available for the centrally planned economies, e.g., China and the U.S.S.R., and is not included in the world production figures shown in Table 4.3. These countries are believed to import a small volume of halomethanes and may also have a small capacity for their production. However, this is not likely to have any significant effect on cumulative world production of halomethanes. Moreover, while there may be a preference for CCl_4 use for grain fumigation in Russia, recent estimates of global emissions of CCl_4 include a portion of this use for grain exported to Russia from the United States.

U.S. Chloromethane Production Locations and Capacities

The companies, locations, and 1976 production capacities of U.S. firms producing chloromethanes are shown in Table 4.4. It should be noted that the numbers on this table are capacities; actual production volumes are not available for each company. Total annual U.S. production figures appear in Table 4.1. Most of the production is in the Gulf Coast region, with lesser amounts in New York, West Virginia, and Kentucky, as shown in Figure 4.3. No information was obtained on locations and capacities for production in other parts of the world.

TABLE 4.3 Estimated 1976 Free World Chloromethane Production

Chloromethane	Regional Production—10^3 Metric Tons (Percent of Total)						
	United States[a]		Europe[b]		Japan[b]		Total
Carbon Tetrachloride	388	(53)	278	(37)	68	(9)	734
Chloroform	133	(50)	104	(40)	26	(10)	263
Methylene Chloride	247	(55)	137	(30)	68	(15)	452
Methyl Chloride	167	(60)	95	(35)	14	(5)	276
Total	935	(54)	614	(35)	176	(10)	1725

[a] U.S. production from Table 4.1.
[b] Derived from Altshuller (1976), Galbally (1976), Singh et al. (1976), and panel estimates.

TABLE 4.4 U.S. Chloromethane Production Locations and Capacities, 1976

		Annual Capacity, 10^3 Metric Tons				
Company	Locations	Carbon Tetrachloride (CCl_4)	Chloroform ($CHCl_3$)	Methylene Chloride (CH_2Cl_2)	Methyl Chloride (CH_3Cl)	Total
Allied Chemical	Moundsville, W. Va.	4	14	23	11	52
Dow Chemical U.S.A.	Freeport, Tex.	61	45	68	—	174
	Pittsburg, Calif.	36	—	—	—	36
	Plaquemine, La.	57	14	86	68	225
E. I. duPont de Nemours	Corpus Christi, Tex.	227	—	—	—	227
FMC (Joint with Allied)	S. Charleston, W. Va.	136	—	—	—	136
Inland Chemical	Manati, P. R.	NA	—	—	—	NA
Stauffer Chemical	LeMoyne, Ala.	91	—	27	—	125
	Louisville, Ky.	16	34	—	7	50
Vulcan Materials	Geismar, La.	41	NA	NA	NA	41
	Wichita, Kans.	27	14	14	NA	55
Diamond Shamrock	Belle, W. Va.	—	8	45	11	46
Continental Oil	Westlake, La.	—	—	—	45	45
Dow Corning	Carrollton, Ky.	—	—	—	9	9
	Midland, Mich.	—	—	—	7	7
Ethyl Corp.	Baton Rouge, La.	—	—	—	45	45
General Electric	Waterford, N.Y.	—	—	—	23	23
Union Carbide	Institute and S. Charleston, W. Va.	—	—	—	23	23
Totals		696	129	263	249	1319

NA: Not available.

SOURCE: Derived from Stanford Research Institute (1976a) and U.S. International Trade Commission (1976a).

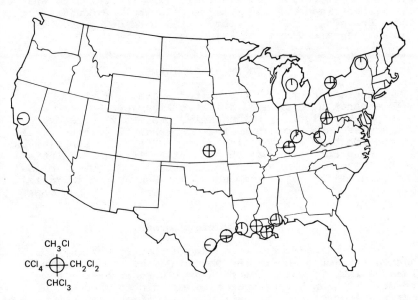

FIGURE 4.3 U.S. chloromethane production locations.[a]

[a]Several of the circles represent more than one of the production locations listed in Table 4.4.

U.S. Production of Bromo-, Iodo-, and Mixed Halomethanes

Published production data are available only for methyl bromide among the compounds which are bromo-, iodo-, and mixed halogenated methanes. These data are shown in Table 4.5, along with production estimates made by chemical industry representatives for the remaining compounds. Only methyl bromide is produced in significant quantities in the United States, the rest of the bromo-, iodo-, and mixed halogenated methanes are small-volume speciality chemicals and their production is relatively minor compared to that of halomethanes made up entirely of chlorine. Production locations and, in the case of methyl bromide, production capacities, are given in Table 4.6.

USES OF HALOMETHANES

End-use patterns in the United States for the major halomethanes have changed dramatically over the years. Carbon tetrachloride, for example, has evolved from dispersive uses accounting for more than 95 percent of all production before 1940 (Altshuller 1976) to consumptive uses accounting for more than 95 percent by 1973 (A.D. Little 1975). The end uses of the halomethanes, discussed below for each of the major compounds, are summarized in Table 4.7.

Carbon Tetrachloride

The principal current use (more than 95 percent) of carbon tetrachloride in the United States is as a chemical intermediate in the production of fluorocarbons. It is reacted with anhydrous hydrofluoric acid to produce the desired degree of fluorination. Approximate consumptions of carbon tetrachloride are 0.60 kilograms per kilogram of fluorocarbon 12 (dichlorodifluoromethane) and 0.53 kilograms per kilogram of fluorocarbon 11 (trichlorofluoromethane) (Stanford Research Institute 1976a). Both fluorocarbons are used in aerosol systems and fluorocarbon 12 also has a major use in refrigeration. Applications of CCl_4 as an industrial solvent, drycleaning agent, fire extinguishing agent, and grain fumigant have decreased in recent years due to its toxicity, a household ban by the FDA, and the availability of chemical alternatives. Substantial amounts exported in recent years have been used for fluorocarbon production in foreign markets.

TABLE 4.5 Approximate U.S. Production of Bromo-, Iodo-, and Mixed Halogenated Methanes, 1975

Compound	Production 10^3 Metric Tons
Methyl Bromide (CH_3Br)	16
Methylene Bromide (CH_2Br_2)	< 2
Bromoform ($CHBr_3$)	< 0.5
Carbon Tetrabromide (CBr_4)	< 0.025
Bromodichloromethane ($BrCHCl_2$)	Lab Quantities
Dibromochloromethane (Br_2CHCl)	Lab Quantities
Methyl Iodide (CH_3I)	< .05
Methylene Iodide (CH_2I_2)	< .05
Iodoform (CH_3I)	< .05
Carbon Tetraiodide (CI_4)	< .05
Total	<19.0

SOURCE: Derived from Stanford Research Institute (1976b) and estimates by chemical industry representatives.

TABLE 4.6 U.S. Production Locations and Capacities for Bromo-, Iodo-, and Mixed Halomethanes, 1976

Company	Location	Annual Capacity, 10^3 Metric Tons						
		Carbon Tetrabromide (CBr_4)	Bromoform ($CHBr_3$)	Methylene Bromide (CH_2Br_2)	Methyl Bromide (CH_3Br)	Iodoform (CHI_3)	Methylene Iodide (CH_2I_2)	Methyl Iodide (CH_3I)
Dow Chemical U.S.A.	Midland, Mich.		X	X	10			
Great Lakes Chemical	El Dorado, Ar.	X			10			
Michigan Chemical	St. Louis, Mich.				2			
Mallinckrodt	St. Louis, Mo.					X		
National Biochemical	Chicago, Ill.						X	X
Sterling Drug, Inc.	Rensselaer, N.Y.						X	
Columbia Organic Chemical	Columbia, S.C.							X
Eastman Kodak	Rochester, N.Y.							X
Fairmont Chemical	Newark, N.J.						X	
RSA Corp.	Ardsley, N.Y.							X
Total					22			

NOTE: X—Capacity data not known for these production sites.
SOURCE: Derived from Chemical Profiles (1976) and Chemical Marketing Reporter (1975).

TABLE 4.7 U.S. End Uses of Major Halomethanes, 1968, 1973 and 1975

Carbon Tetrachloride (CCl$_4$)

Production, 10^3 metric tons (%)

End Use:	1968		1973	
Fluorocarbon 12 production	197	(57)	285	(60)
Fluorocarbon 11 production	107	(31)	166	(35)
Other (fumigant, solvent)	42	(12)	24	(5)
Total	346	(100)	475	(100)

Chloroform (CHCl$_3$)

Production, 10^3 metric tons (%)

End Use:	1968		1973	
Fluorocarbon 22 (gas) production	49	(60)	60	(52)
Fluorocarbon 22 (for plastic) production	25	(30)	47	(41)
Other	8	(10)	8	(7)
Total	82	(100)	115	(100)

Methylene Chloride (CH$_2$Cl$_2$)

Production, 10^3 metric tons (%)

End Use:	1968		1973	
Paint Remover	34	(25)	94	(40)
Solvent degreasing	27	(20)	24	(10)
Aerosol sprays (solvent, vapor pressure depressant)	14	(10)	19	(8)
Plastics processing	16	(12)	14	(6)
Other (coatings, extraction solvent)	45	(33)	38	(16)
Export	--	--	47	(20)
Total	136	(100)	236	(100)

Methyl Chloride (CH$_3$Cl)

Production, 10^3 metric tons (%)

End Use:	1968		1973	
Silicone production	83	(60)	106	(43)
Tetramethyl lead production	25	(18)	94	(38)
Butyl rubber production	17	(12)	10	(4)
Methyl cellulose production	8	(6)	7	(3)
Herbicide production	1	(1)	5	(2)
Quaternary amine production	1	(1)	7	(3)
Other (including export)	3	(2)	17	(7)
Total	138	(100)	246	(100)

Methyl Bromide (CH$_3$Br)

Production, 10^3 metric tons (%)

End Use:	1975	
Soil Fumigant	9	(55)
Space Fumigant	2	(15)
Other	1	(5)
Export (mostly for soil fumigation)	4	(25)
Total	16	(100)

Other Bromo- and Iodo- Methanes

Compounds		End Use:
CH$_2$Br$_2$	(Methylene bromide)	Chemical intermediate
CHBr$_3$	(Bromoform)	Chemical intermediate
CBr$_4$	(Carbon tetrabromide)	Chemical intermediate
BrCHCl$_2$	(Bromodichloromethane)	Research
Br$_2$CHCl	(Dibromochloromethane)	Research
CH$_3$I	(Methyl iodide)	Pharmaceutical intermediate
CH$_2$I$_2$	(Methylene iodide)	Research
CHI$_3$	(Iodoform)	Research
CI$_4$	(Carbon tetraiodide)	Research

SOURCE: Derived from Chemical Marketing Reporter (1975), Stanford Research Institute (1976b), and A. D. Little (1975).

Chloroform

Chloroform, like carbon tetrachloride, is used in the United States primarily for the production of fluorocarbons, principally fluorocarbon 22 (chlorodifluoromethane), which is largely used as a refrigerant. Approximate consumption of chloroform for this use, assuming similar yields as for the other fluorocarbons, is 0.65 kilograms per kilogram of fluorocarbon 22. Other uses of chloroform are as a chemical intermediate in the production of dyes and pesticides, and as a solvent and/or formulating agent in certain pharmaceutical products. Chloroform is no longer used as a general anesthetic for humans and recently the FDA took action to eliminate it from pharmaceutical products.

Methylene Chloride

Methylene chloride is the most diversely used chloromethane in the United States although it ranks behind carbon tetrachloride in total production volume. Almost all uses are dispersive, i.e., the compound is lost directly to the environment through use. Most of its applications make use of its property as an excellent solvent for both polar and nonpolar materials. In addition, the relatively low boiling point and absence of a flash point make methylene chloride useful in aerosol systems and facilitate its removal in certain solvent applications. Good solvent properties and low toxicity have made it particularly useful in pharmaceutical and food extraction applications at low temperatures. It is commonly used in paint stripping and degreasing operations.

Methyl Chloride

Methyl chloride is used primarily in the United States as a chemical intermediate. It is used as a methylating agent in the production of silicones, tetramethyl lead, methyl cellulose, quaternary ammonium compounds, and a variety of other compounds. There was almost a two-fold increase in tetramethyl lead end use from 1968 to 1973; however, this use probably has declined or held constant since 1973 due to greater use of nonleaded gasoline. The "other" category in Table 4.7 includes use as a blowing agent in plastic foams.

Methyl Bromide

Methyl bromide use in the United States includes its employment as a soil fumigant for tomatoes, strawberries, tobacco, ornamentals, and other crops. It is no longer used as a fire extinguishing agent. Methyl bromide production has grown at the rate of 7.5 percent in the United States, with an anticipated future growth of 7 percent, mostly for the export market (Chemical Marketing Reporter 1975).

Other Halomethanes

The other halomethanes of interest are used in the United States as chemical intermediates. Since these materials are quite expensive, uses are limited to those applications where other compounds are not suitable. Examples are the use of methyl iodide for quaternization of highly hindered amines and the use of bromoform and carbon tetrabromide as chain transfer agents in the preparation of certain polymers. In most of these applications, the halogen is converted to the inorganic form before entry into the environment, or in many cases is recycled to recover the halogen. There are no known industrial applications for several of the compounds listed.

REFERENCES

Altshuller, A.P. (1976) Average tropospheric concentration of carbon tetrachloride based on industrial production, usage, and emissions. Environmental Science and Technology 10:596-8.

Arthur D. Little, Inc. (1975) Preliminary Economic Impact Assessment of Possible Regulatory Action to Control Atmospheric Emissions of Selected Halocarbons. Prepared by A.D. Little, Inc., EPA-450/3-75-073. NTIS No. PB-247 115. Springfield, Va.: National Technical Information Service.

Chemical Marketing Reporter (1975) Chemical Profile, Methyl Bromide. Chemical Marketing Reporter 208(7):8-9.

Chemical Profiles (1976) Methyl bromide. New York: Schnell Publishing Co.

Galbally, I.E. (1976) Man-made carbon tetrachloride in the atmosphere. Science 193:573-6.

McCarthy, R.L. (1974) Fluorocarbons in the Environment. Paper presented before the National Geophysical Union, San Francisco, California, December 31, 1974. Wilmington, Del.: E.I. du Pont de Nemours & Co.

Singh, H.B., D.P. Fowler, and T.O. Peyton (1976) Atmospheric carbon tetrachloride: another man-made pollutant. Science 192:1231-4.

Stanford Research Institute (1976a) Chloromethanes - Salient Statistics. In Chemical Economics Handbook, sections 635.2030 (carbon tetrachloride), 635.5030 (chloroform), 635.6030 (methyl chloride), 635.6040 (methylene chloride); April 1976 et seq. Menlo Park, Calif.: Stanford Research Institute. (This is a client-private service of SRI's Chemical Information Services Department, available by subscription only.)

Stanford Research Institute (1976b) In Chemical Economics Handbook, section 573.9005 (methyl bromide); December 1976. Menlo Park, Calif.: Stanford Research Institute. (This is a client-private service of SRI's Chemical Information Services Department, available by subscription only.)

U.S. International Trade Commission (1976) Preliminary Report on U.S. Production of Selected Synthetic Organic Chemicals, May, June, and Cumulative Totals, 1976. Synthetic Organic Chemicals. Series C/P-76-5, August 18, 1976. Washington, D.C.: International Trade Commission.

CHAPTER 5

ENVIRONMENTAL OCCURRENCE, TRANSPORT, AND FATE

SOURCES OF HALOMETHANES TO THE ENVIRONMENT

In its study of sources, the Panel considered halomethanes under three general categories: those emitted in product manufacture, distribution, and use; those that are secondary emissions or formations; and those that have primarily natural rather than anthropogenic origins.

Emissions from Product Manufacture and Use

This category of halomethane losses includes losses that occur in the production, transport, and storage of manufactured products; losses when compounds are "consumed" during use as a chemical intermediate in the production of other chemicals; losses when halomethanes are present as impurities in other products; and dispersive losses resulting from the end use of manufactured products.

The estimated 1973 U.S. emissions from these sources are summarized in Table 5.1. Approximately 77 percent by weight of all annual emissions are estimated to originate from the dispersive uses of methylene chloride. The next major contributions to total emissions come from carbon tetrachloride and methyl bromide.

Direct Losses from Production, Transport, and Storage

Halomethane losses during production, transport, and storage are not well documented, but it appears that such losses represent only about 5 percent of all halomethanes entering the environment from product manufacture and use (see Table 5.1). Most halomethane losses in production, transportation, and storage are the fugitive type, i.e., transient releases due to leaky pump seals, valves, and joints. Some come directly from vents while others may occur as a result of interruptions in normal plant operations, equipment malfunctions, normal maintenance, and in transfer during distribution.

TABLE 5.1 Estimated Annual U.S. Losses of Nonfluorinated Halomethanes, 1973[a]

Compound	Production, Transport, & Storage	Consumptive Losses (Chem. Intermediate)	Impurities in Other Products	Dispersive Losses in Use of Product	Total	Losses as Percent of Each Compound Produced	Percent of All Compound Losses
				(in 10^3 metric tons)			
Chlorinated Methanes							
CCl_4	7.1	4.5	0.4	22.6	34.6	7.3	13.5
$CHCl_3$	1.7	1.1	0.3	3.4	6.5	5.6	2.5
CH_2Cl_2	2.4	0	0.3	196.0	198.7	84.2	77.5
CH_3Cl	1.7	1.4	0.4	1.7	5.2	2.1	2.0
Brominated Methanes							
CBr_4	<.0001	<.0001	0	0	<.0002		
$CHBr_3$	<.003	<.003	0	0	<.005		
CH_2Br_2	<.01	<.01	0	0	<.02		
CH_3Br	0.1	0	0	11.3	11.4	71.0	4.5
Total (all compounds)	13.0	7.0	1.4	235.0	256.4	24.0[b]	100.0
Percent (by type of loss)	5.0	2.7	.5	91.8	100.0		

[a] Iodinated and mixed halomethanes do not appear because production volumes are so low or non-existent, that losses are nil.
[b] Percentage loss of all compounds produced (1.07×10^6 metric tons in 1973).

SOURCES: Data on emissions from production, distribution, and storage for the chloromethanes from A.D. Little, Inc. (1975). Similar assumptions were used on percentages lost for the other compounds. Dispersive losses for CCl_4 were modified by the panel.

Estimates made by A.D. Little, Inc. (1975) of U.S. chloromethane losses from production, transport, and storage range from 0.7 percent to 1.5 percent of their total production volume. The lowest value is assigned to the most volatile compound, methyl chloride, since special handling procedures are required to prevent major losses. The geographic origins of production losses of chloromethanes can be estimated from data provided in Table 4.4 for the locations shown on Figure 4.3.

Relatively little information is available on halomethane spills, making it difficult to judge the quantity of emissions from spills. However, a number of spills have been recorded. U.S. Coast Guard records show a spill of 8,327 liters of methyl chloride in June 1973 into Chesapeake Bay; a spill of 1892 liters of carbon tetrachloride into the Kanawha River, West Virginia, in April 1975; and a spill of 242,240 liters of chloroform in September 1973 into the lower Mississippi.[1]

A second massive spill of 1.75 million pounds of chloroform into the Mississippi occurred when two barge tanks ruptured at Baton Rouge, Louisiana. The event is of particular interest, not only as an example of such a spill in a river and the concentrations of chloroform that can result, but also because the results were reported and studied in some detail (Neely et al. 1976).

Concentrations up to approximately 300 parts per billion (ppb) (µg/l) were measured at two points 16.3 and 121 miles downstream, up to 180 hours after the spill. A model was developed and fit to the concentration-time profile; the results for the 16.3 mile point are shown in Figure 5.1.

The characteristics of this model are of interest because of its possible use in predicting the behavior of similar halogenated hydrocarbons in natural water systems. The model "precludes the existence of a dynamic equilibrium between the chloroform and mud in the river bottom," and cites supporting evidence (Neely et al. 1976:75). The model shows the existence of a layer of chloroform at the river bottom resulting from its high density and low water solubility; exchange between this bottom layer and the boundary layer above it; exchange between the boundary layer and the soluble chloroform in the flowing river compartment above; and evaporation from the river compartment into the atmosphere.

Two important conclusions of the study are that after the spill most of the chloroform went into the bottom layer, and that diffusion from this layer into the boundary layer was considerably slower than transport from the boundary layer into the bulk flowing compartment.

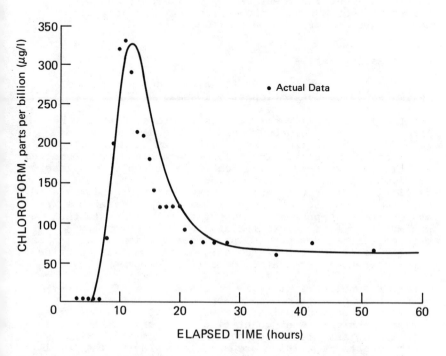

SOURCE: Neely et al. (1976) Mathematical models predict concentration-time profiles resulting from chemical spill in a river. Reprinted with permission from *Environmental Science and Technology* 10:72-76. Copyright by the American Chemical Society.

FIGURE 5.1 Concentration-time profile of chloroform in a river (16.3 miles from point of addition).

Consumptive Losses

Consumptive losses of halomethanes used as chemical intermediates in the production of other chemicals are estimated to be similar in magnitude to direct losses during production, transport and storage of the original halomethanes (A.D. Little, Inc. 1975). Consumptive losses represent about 3 percent of the estimated amount of chloromethanes entering the environment from product manufacture and use.

Losses as Impurities

Less than 1 percent of all halomethane losses are estimated to be due to impurity losses.

Low concentrations of various chloromethanes may be present in other chlorocarbons, but the amounts are relatively insignificant even for the major high-volume chlorinated products. The chlorocarbon products considered in this respect were: vinyl chloride, perchloroethylene, trichloroethylene, 1,1,1-trichloroethane, ethylene dichloride, fluorocarbons, chlorine, and hydrochloric acid. Chloromethanes are not listed in sales specifications for any of these compounds. Industry sources indicate that the maximum levels of chloromethanes in these commodity chemicals rarely exceed 500 parts per million (ppm) and are usually below 100 ppm. The residual carbon tetrachloride in fluorocarbons is estimated to be below 2 ppm.[2]

Best Panel estimates are that a maximum of 400 metric tons a year of each of the chloromethanes enter the environment as impurities. Bromo- and iodomethanes are not present in significant quantities as impurities in elemental chlorine or any of the compounds made from it. Overall, losses due to impurities are small compared to the total amounts of halomethanes entering the environment from other sources.

Dispersive Uses

For all major halomethanes, dispersive uses are the main routes of entry into the environment from product manufacture and use, accounting for more than 90 percent of all estimated losses. These losses are largely by direct emissions to the atmosphere. Where compounds are used as extraction solvents, the potential also exists for minor discharges to water.

The halomethane with the greatest dispersive loss is methylene chloride (CH_2Cl_2). The loss amounts to 84 percent of all CH_2Cl_2 production and approximately 77 percent of all halomethane emissions. The dispersive uses of CH_2Cl_2 are varied and widespread and are distributed geographically approximately with the industrialized population in the United States. Although most of the losses are to the atmosphere, CH_2Cl_2 is relatively soluble and, when it is used as an extraction solvent, the potential exists for concentrations up to 1,500 ppm to occur in water effluents depending on initial quantities used, the purpose and design of the extraction process, and the nature of terminal treatment applied (see Table 5.2). The most effective means of removal of CH_2Cl_2 from water is air stripping, which hastens the transfer of the compound from water to the atmosphere.

Since use of carbon tetrachloride (CCl_4) as an industrial solvent, drycleaning agent, and fire extinguishing agent has been largely discontinued, the major dispersive use of CCl_4 is as a component of liquid grain fumigants in marine transportation. Some losses occur when CCl_4 is pumped into the hold of ships, some during transit, some as a result of chemical decomposition, and some when grain is unloaded and further distributed (Kenaga 1957). A major loading point for grain is New Orleans with variable international destinations. Other dispersive uses of CCl_4 are minor and have declined in recent years due to the increased recognition of CCl_4 toxicity.

Dispersive uses of chloroform ($CHCl_3$) are largely in a variety of pharmaceutical formulation processes. Emissions are likely to be both to water and the atmosphere at pharmaceutical plant sites.

Dispersive uses of methyl chloride are small (see Table 5.1) and can account for only a small percent of the methyl chloride found in the atmosphere. The major input of methyl chloride to the environment is thought to be from natural processes. Because of its high vapor pressure, anthropogenic emissions of methyl chloride are almost entirely to the atmosphere.

Soil fumigation is the major use for methyl bromide (CH_3Br) (Getzendaner and Richardson 1966, Getzendaner et al. 1968). The compound is either injected under the surface of the soil or applied to the soil surface under tarps. Although a small amount probably is lost to the atmosphere during these applications, most of the CH_3Br is trapped in the soil. CH_3Br as a soil fumigant is chiefly used for tomato, strawberry, tobacco, and ornamental crops. Its major users are California farmers.

Space fumigation with CH_3Br is carried out on a variety of processed foods, grains, and nuts (Kenaga 1957). The storage warehouses, where fumigation usually is done, may be located near agricultural production centers, distribution centers, or population centers. Much of the compound is converted to inorganic bromide during fumigation, probably by reaction with amino or sulfhydryl groups in contacted proteins. An estimated 30 percent of the CH_3Br used for this purpose may escape from the fumigation chamber and enter the environment while the rest decomposes to inorganic bromide and methylated derivatives of organic compounds.

Secondary Emissions and
Secondary Formation from Anthropogenic Sources

The second major category of anthropogenic emissions includes those produced through secondary formation reactions or as incidental by-products of anthropogenic activities. Emissions resulting from the use of chlorine to treat municipal water supplies and from the bleaching of paper pulp are considered in this section, as are emissions from combustion processes and thermal degradation. Although some of the emissions in industrial wastewater effluents were considered under production losses, the Panel decided to include them here also because of similarities in data on secondary formation from chlorine used in industrial processes and municipal water supply treatment, and primary losses in industrial and municipal liquid wastes.

Halomethane Generation in Chlorination Processes

In 1971 in the United States alone, the estimated production of chlorine was about 8.4×10^6 metric tons, of which about 4 percent was used for municipal water and wastewater treatment, and 15 percent was used by the pulp and paper industry (Hampel and Hawley 1973). Thus, the potential for forming considerable quantities of halomethanes in these uses is great.

Measurements of halomethanes emitted as a result of the use of chlorine in industrial processes and in treatment of municipal water supplies are shown in Tables 5.2 through 5.6. (For a discussion of trihalomethane formation in municipal drinking waters see the section of this Chapter on Finished Drinking Water.) In some cases, these data show considerable concentrations of a few halomethanes. Of particular interest is the presence of methylene chloride and chloroform in paper mill effluents, with concentrations as high as 42,000 (Table 5.2) and 1,700 µg/l (Table 5.5), respectively. It is likely that concentrations had been even higher, but were reduced by the volatile action of these compounds. These halomethanes probably are a secondary formation, whereas the concentrations of methylene

TABLE 5.2 Concentrations of Methylene Chloride (CH_2Cl_2) in Wastewater Effluents

Source	Concentration ($\mu g/l$)
North Side Sewage, Chicago, Ill.	>15
West Side Sewage, Chicago, Ill	8
Calumet Sewage, Chicago, Ill	3
Rubber companies (latex), Louisville and Calvert City, Ky.	1,200-11,000
McAlpine Creek Sewage Plant, Upper Catawba River, N.C.	10
Pharmaceutical company, Pacolet and Enoree Rivers, S.C.	17
Paper Mill No. 1, Mobile River, Ala.	42,000
Paper Mill No. 2, Mobile River, Ala.	40,000
Paper Mill No. 3, Mobile River, Ala.	3,800- 9,600
Paper Mill No. 4, Mobile River, Ala.	19,000
Electrical manufacturing, Mt. Vernon, Ind.	132,000
Chemical company, Pascolet and Enoree Rivers, S.C.	4-200
Chemical company, Louisville, Ky	1,600-43,000
Dalton Sewage, Dalton, Ga.	100
Calhoun Sewage, Calhoun, Ga.	5
Rome Sewage, Rome, Ga.	290

SOURCE: Personal communication from Larry Keith (1976) Environmental Research Laboratory, U.S. EPA, Athens, Georgia.

TABLE 5.3 Concentrations of Chlorodibromomethane ($ClCHBr_2$) in Wastewater Effluents

Source	Concentration ($\mu g/l$)
Chemical company, Europe	650
Chemical company, Calvert City, Ky.	10-100
West Side Sewage, Chicago, Ill.	0.05
North Side Sewage, Chicago, Ill.	1

SOURCE: Personal communication from Larry Keith (1976) Environmental Research Laboratory, U.S. EPA, Athens, Georgia.

TABLE 5.4 Concentrations of Carbon Tetrachloride (CCl_4) in Wastewater Effluents

Source	Concentration ($\mu g/l$)
Rubber company (latex), Calvert City, Ky.	10-100
Chemical companies, Calvert City and Louisville, Ky.	10-5,000

SOURCE: Personal communication from Larry Keith (1976) Environmental Research Laboratory, U.S. EPA, Athens, Georgia.

TABLE 5.5 Concentrations of Chloroform ($CHCl_3$) in Wastewater Effluents

Source	Concentration ($\mu g/l$)
Paper mill aerated lagoon, Springfield, Ore.	10-20
North Side Sewage, Chicago, Ill.	>20
West Side Sewage, Chicago, Ill.	5
Calumet Sewage, Chicago, Ill.	14-20
Rubber companies, Louisville and Calvert City, Ky.	2,600-22,000
Chemical companies, Louisville and Calvert City, Ky.	10-20,000
Paper Mill No. 3, Mobile River, Ala.	270-1,700
Crude sewage, Europe	9
Biological treatment plant effluent, Europe	7
Chlorinated biological treated effluent, Europe	12

SOURCE: Personal communication from Larry Keith (1976) Environmental Research Laboratory, U.S. EPA, Athens, Georgia.

TABLE 5.6 Concentrations of Various Halocarbons in Wastewater Effluents

Compounds	Source	Concentration ($\mu g/l$)
Bromoform ($CHBr_3$)	Chemical company, Europe	4,000
	Chemical company, Calvert City, Ky.	10-100
Methylene Bromide (CH_2Br_2)	Chemical company, Europe	200
Methyl Chloride (CH_3Cl)	Chemical company, Louisville, Ky.	1,000-10,000
	West Side Sewage, Chicago, Ill.	1
Bromodichloromethane ($BrCHCl_2$)	Chemical companies, Louisville and Calvert City, Ky.	10-10,000

SOURCE: Personal Communication from Larry Keith (1976) Environmental Research Laboratory, U.S. EPA, Athens, Georgia.

chloride and chlorodibromomethane (Tables 5.2 and 5.3) in effluent from the rubber company in Kentucky and the chemical companies probably are primary and stem from direct use of the compounds in the chemical process.

Although information is insufficient for judging the quantity or net impact of halomethanes in liquid industrial and municipal waste effluents, these data indicate that in local situations there can be relatively large emissions that might create high concentrations in the immediate receiving waters. One concern is the possible effect on downstream water supplies. A second involves the possibility of elevated ambient air concentrations of these compounds in the vicinity of liquid effluents. Where lagooning is practiced before treatment or release of the effluents, it is likely that a large fraction of the compounds will be released to the air before the liquids reach the receiving stream.

The data in Tables 5.2 through 5.6 show that a variety of industrial sources produce significant concentrations of the halomethanes. Thus, if decisions are contemplated for the control of these compounds in local surface waters, not only should their generation in chlorination processes be evaluated, but also the likelihood of emissions from local industrial sources.

Combustion Processes and Thermal Degradation

Thermal decomposition of plastics, degradation of ethylene dibromide in the combustion of gasoline, and cigarette smoke have been identified as sources of halomethanes to the environment; forest fires and agricultural burning as a source of methyl chloride and methylene chloride are discussed later in this chapter under natural sources. However, the quantitative contribution of halomethanes from these sources cannot be evaluated. All of these combustion processes represent non-continuous point sources and their significance probably lies in local or isolated human exposures at concentrations above ambient levels.

Several materials have been identified as having low molecular weight halogenated hydrocarbons, especially methyl chloride, as products of their thermal decomposition. In general, these products form by the reaction of flame retardants added to the material and organics that are in the material or produced in the decomposition. Quantitative measurements of these products are only estimates and the strength of this source cannot be evaluated.

Paciorek et al. (1974b) and Derby and Freedman (1974) have both found evidence that methyl chloride is a decomposition product of polyvinyl chloride (PVC). The latter investigators identified methyl chloride as a vapor-phase pyrolysis product of pure PVC by gas chromatography. Paciorek and coworkers conducted sealed tube studies (static) and stagnation burner studies (dynamic) on the the oxidative thermal decomposition of several PVC compositions. The amount of methyl chloride that was produced depended upon the PVC composition and the type of oxidation process. The range of production was 0.13 to 3.75 mg methyl chloride per gram of PVC. For each composition, less methyl chloride was produced in the stagnation burner than in the sealed tube.

Thrune (1963) identified methyl chloride, methyl bromide, and methylene bromide as minor decomposition products when DER 542 resin is burned, and methyl chloride as a decomposition product when X-3448, Dow epoxy resin, is burned. DER 542 is the diglycidal ether of tetrabromobisphenol-A, which was cured with methylene dianiline (MDA) for this experiment, and X-3448 is the diglycidal ether of tetrachlorobisphenol-A, which was cured with methylene dianiline (MDA) for this experiment. Neither resin produced any of the low molecular weight halogenated hydrocarbons when cured with methyl nadic anhydride (MNA). The following quantitative approximations were made on the basis of the laboratory experiment:

Product	Resins	
	DER 542 (MDA)	X-3448 (MDA)
CH_3Br (x 10^{-4} moles of gas per 10 gm resin burned)	0.047	---
CH_3Cl (x 10^{-4} moles of gas per 10 gm resin burned)	0.007	0.008
CH_2Br_2 (x 10^{-4} moles of gas per 10 gm resin burned)	0.004	---

Paciorek et al. (1974a) examined oxidative thermal decomposition products of selected polymers used in underground mines. They found that jute, which is used extensively in mines in brattice cloths, forms methyl chloride as a thermal decomposition product at 475-555°C.

The gaseous samples contained between 0.07 and 0.40 mole percent methyl chloride.

The presence of methyl chloride in cigarette smoke has been reported by several authors. Johnson et al. (1973) identified methyl chloride as a constituent of the mainstream smoke (emitted through the butt end of the cigarette during puffing) and the sidestream smoke (emitted between puffs while the cigarette smoulders). Their findings indicate that more methyl chloride is produced while a cigarette smoulders than during actual puffing.

Philippe and Hobbs (1956) and Newsome et al. (1965) identified methyl chloride as a component of the gas phase of cigarette smoke. The former authors report a quantitative value of 3.2×10^{-2} ml of methyl chloride per puff of a cigarette at normal temperature and pressure.

Chopra and Sherman (1972) identified both methyl chloride and chloroform in the smoke of cigarettes in which tobacco fumigated with p,p'-DDT was used (tobacco is no longer treated with p,p'-DDT in the United States and some other countries in the northern hemisphere where its use has been banned). Their experiments indicate a linear relationship between the amount of chloroform that is formed and the amount of p,p'-DDT in the tobacco. However, large amounts of methyl chloride are formed from the burning of tobacco, regardless of whether tobacco is treated with p,p'-DDT. Methyl chloride probably forms by the action of methyl free radicals in tobacco smoke on organic and inorganic chlorine in tobacco. The investigators note that their findings of the amount of methyl chloride formed per gram of tobacco smoked is comparable to the amount reported by Philippe and Hobbs. On the basis of eight puffs per cigarette, the Chopra and Sherman findings of the amount of methyl chloride per cigarette smoked translate to 11.6 mg of methyl chloride per 20-cigarette pack smoked. They also found that no methylene chloride forms.

The Chopra and Sherman findings indicate that methyl chloride is produced mainly from tobacco inorganic chlorine reactions, which occur during smoking, at a rate of 2.0 mg/g of tobacco. On the basis of 1974 world tobacco production of 5.23×10^6 tons/yr (United Nations Monthly Bulletin of Statistics, June 1975), the dispersion of methyl chloride to the atmosphere from cigarette smoking is 1.05×10^{10} g/yr. This input is clearly less than the estimated natural production of 5.2×10^{12} g/yr (Yung et al. 1975) and an estimated anthropogenic input of 2×10^{10} g/yr (this study). Thus, tobacco smoking appears to add an insignificant amount of methyl chloride to the global atmosphere. However, in buildings and homes and in highly populated areas, the

methyl chloride content of the interior atmosphere could in
principle be governed by the production of cigarette smoke.

Other studies of halomethanes produced by combustion
include Backus,[3] which listed chloroform as a volatile
degradation product detected in the pyrolysis of rigid
urethane foams at 500°C.

Methyl bromide, in addition to being emitted directly as
a fumigant, is a secondary formation product from ethylene
dibromide (CH_2Br-CH_2Br) in the combustion of gasoline
(Harsch and Rasmussen 1977; see also Analysis of Global Mass
Balances, this chapter). Most anthropogenic bromine goes
into the formation of the gasoline additive, ethylene
dibromide.

Atmospheric Conversions

Several atmospheric conversions that result in the
formation of halomethanes have been postulated. Further
study is needed, however, to determine the significance of
these reactions in the real atmosphere. These include the
conversion of ethylene dichloride and vinyl chloride to
methyl chloride; the photooxidation of trichloroethylene to
chloroform; and the formation of carbon tetrachloride from a
perchloroethylene precursor. These conversions are
discussed later in this chapter in the Analysis of Global
Mass Balances.

When methyl bromide is used as a fumigant, a minor
secondary formation of methyl chloride takes place under
closed conditions in certain food commodities. Dennis et
al. (1972) say that the presence of methyl chloride during
such fumigations probably would not be detected by the
methods generally used for methyl bromide analyses. They
also note that the rate of generation and concentration of
methyl chloride gas probably depends on the same factors
that influence methyl bromide concentrations, such as
moisture content, temperature, the concentration applied,
type of product, and amount of milling.

Summary of Emissions from Anthropogenic Sources

Carbon Tetrachloride

Annual and cumulative U.S. emissions of carbon
tetrachloride (CCl_4) are of interest, because this compound
appears to be stable in the troposphere. Reported estimates
are affected by two assumptions: the relative percentages
of total annual production used for dispersive and
consumptive uses, and the percentages actually lost in each

of these uses. Basic production figures employed by several
authors (Altshuller 1976, Galbally 1976, Singh et al. 1976)
and those used in this study to calculate total CCl_4
emissions are essentially the same.

Panel estimates of United States and "free world"
anthropogenic emissions of CCl_4 for almost 70 years are
summarized in Figures 5.2 and 5.3. Other known sources of
CCl_4 emissions are insignificant. Most CCl_4 emissions
originated in the United States with minor fluctuations
occurring prior to about 1922 and somewhat wider swings
taking place in the years 1940 to 1960. The drop in annual
U.S. emissions of carbon tetrachloride since 1970 can be
explained by the fact that production leveled off beginning
in 1970 and dispersive uses continued to fall, eventually
leveling off at about 5 percent of total production. It is
estimated that current U.S. emissions are about 28×10^3
metric tons per year.

Western Europe and Japan seem to account for most of the
known non-U.S. emissions. Emissions from the countries with
centrally planned economies, such as China and the U.S.S.R.,
are hard to estimate since no production and use data are
available. However, cumulative emissions from Eastern
Europe, Australia, South America, Africa, and Asia have been
estimated at about 10 percent of those in the United States,
Western Europe, and Japan (Singh et al. 1976). The pattern
of "free world" emissions is similar to that in the United
States, but again it should be emphasized that uncertainties
exist in computing overall estimates of this type. The
fitted curves in Figures 5.2 and 5.3 were based on
calculations made during the present study. The annual and
cumulative emission components are presented in Table 5.7.
The basic production data were obtained from U.S.
International Trade Commission reports and other historical
records. Dispersive and consumptive use fractions of total
production were estimated from the work of several authors
(A.D. Little, Inc. 1975, Altshuller 1976, Galbally 1976,
Singh et al. 1976). Losses of CCl_4 during production were
assumed linear from 5 percent in 1910 decreasing to 1.5
percent in 1973 (Singh et al. 1976). Dispersive losses were
estimated to be 95 percent of all dispersive uses
(Altshuller 1976). Consumptive losses were estimated to be
1.0 percent of all consumptive uses (A.D. Little, Inc.
1975). Non-U.S. production and losses were based on
estimates by Singh et al. (1976), following a review of
Altshuller, Galbally, and Singh data.

Table 5.8 summarizes annual and cumulative emissions of
carbon tetrachloride as reported by various authors.
Although the authors differ in the amount of annual
emissions they estimate for a given year, their studies as a
whole show that total annual emissions of CCl_4 have leveled

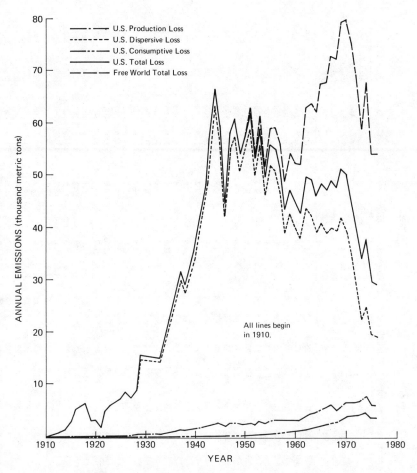

SOURCE: Panel estimates based on U.S. International Trade Commission and other historical records (production); A. D. Little, Inc. (1975), Altshuller (1976), Singh et al. (1976), Galbally (1976) (dispersive and consumptive use of fractions); Singh et al. (1976) (non-U.S. production and losses); and other research used in this study.

FIGURE 5.2 Annual U.S. and "Free World" total emissions of carbon tetrachloride, 1910-1976.

TABLE 5.7 Annual and Cumulative U.S. and "Free World" Emissions of Carbon Tetrachloride (in 1000 metric tons)

Year	Annual U.S. Prod. KT	Prod. Loss %	Prod. Loss KT	Dis. Use %	Dis. Use KT	Dis. Loss %	Dis. Loss KT	Con. Use %	Con. Use KT	Con. Loss %	Con. Loss KT	*Annual U.S. Emis. KT	Cum. U.S. Emis. KT	Annual U.S./World %	Annual "Free World" Emis. KT	Cum. "Free World" Emis. KT
1908	0.1	5.0	0.0	100	0.1	95	0.1	0	—	1.0	—	0.1	0.1	100	0.1	0.1
1909	0.3	5.0	0.0	100	0.3	95	0.3	0	—	1.0	—	0.3	0.4	100	0.3	0.4
1910	0.2	5.0	0.0	100	0.2	95	0.2	0	—	1.0	—	0.2	0.6	100	0.2	0.6
1911	0.7	5.0	0.0	100	0.7	95	0.6	0	—	1.0	—	0.7	1.3	100	0.7	1.3
1912	0.8	5.0	0.0	100	0.8	95	0.7	0	—	1.0	—	0.8	2.0	100	0.8	2.0
1913	1.0	5.0	0.1	100	1.0	95	1.0	0	—	1.0	—	1.0	3.1	100	1.0	3.1
1914	1.5	5.0	0.1	100	1.5	95	1.4	0	—	1.0	—	1.5	4.5	100	1.5	4.5
1915	2.8	5.0	0.1	100	2.8	95	2.6	0	—	1.0	—	2.8	7.3	100	2.8	7.3
1916	5.1	5.0	0.3	100	5.1	95	4.8	0	—	1.0	—	5.1	12.4	100	5.1	12.4
1917	5.9	4.5	0.3	100	5.9	95	5.6	0	—	1.0	—	5.9	18.3	100	5.9	18.3
1918	6.4	4.5	0.3	100	6.4	95	6.1	0	—	1.0	—	6.4	24.7	100	6.4	24.7
1919	3.3	4.5	0.1	100	3.3	95	3.1	0	—	1.0	—	3.2	27.9	100	3.2	27.9
1920	3.2	4.5	0.1	100	3.2	95	3.0	0	—	1.0	—	3.2	31.1	100	3.2	31.1
1921	1.9	4.5	0.1	100	1.9	95	1.8	0	—	1.0	—	1.8	32.9	100	1.8	32.9
1922	5.1	4.5	0.2	100	5.1	95	4.8	0	—	1.0	—	5.1	38.0	100	5.1	38.0
1923	6.1	4.5	0.3	100	6.1	95	5.8	0	—	1.0	—	6.1	44.1	100	6.1	44.1
1924	6.5	4.5	0.3	100	6.5	95	6.2	0	—	1.0	—	6.5	50.5	100	6.5	50.5
1925	7.3	4.0	0.3	100	7.3	95	7.0	0	—	1.0	—	7.3	57.8	100	7.3	57.8
1926	8.6	4.0	0.3	100	8.6	95	8.2	0	—	1.0	—	8.5	66.3	100	8.5	66.3
1927	7.5	4.0	0.3	100	7.5	95	7.2	0	—	1.0	—	7.5	73.8	100	7.5	73.8
1928	8.9	4.0	0.4	100	8.9	95	8.5	0	—	1.0	—	8.8	82.6	100	8.8	82.6
1929	15.7	4.0	0.6	100	15.7	95	15.0	0	—	1.0	—	15.6	98.2	100	15.6	98.2
1930	15.6	4.0	0.6	100	15.6	95	14.8	0	—	1.0	—	15.4	113.6	100	15.4	113.6
1931	15.8	4.0	0.6	97	15.3	95	14.5	3	0.5	1.0	0.0	15.2	128.8	100	15.2	128.8
1932	16.1	4.0	0.6	95	15.3	95	14.5	5	0.8	1.0	0.0	15.2	144.0	100	15.2	144.0
1933	16.2	4.0	0.6	93	15.1	95	14.3	7	1.1	1.0	0.0	15.0	158.9	100	15.0	158.9
1934	21.3	3.5	0.7	91	19.4	95	18.4	9	1.9	1.0	0.0	19.2	178.1	100	19.2	178.1
1935	25.3	3.5	0.9	89	22.5	95	21.4	11	2.8	1.0	0.0	22.3	200.4	100	22.3	200.4
1936	31.1	3.5	1.1	87	27.0	95	25.7	13	4.0	1.0	0.0	26.8	227.2	100	26.8	227.2
1937	38.1	3.5	1.3	84	32.0	95	30.4	16	6.1	1.0	0.	31.8	259.0	100	31.8	259.0
1938	35.4	3.5	1.2	82	29.0	95	27.6	18	6.4	1.0	0.	28.9	287.9	100	28.9	287.9
1939	41.1	3.5	1.4	80	32.8	95	31.2	20	8.2	1.0	0.	32.7	320.6	100	32.7	320.6

Year																
1940	45.7	3.5	1.6	78	35.7	33.9	22	10.1	1.0	0.1	35.6	100	356.2	35.6	356.2
1941	55.2	3.5	1.9	76	42.0	39.9	24	13.3	1.0	0.1	42.0	100	398.2	42.0	398.2
1942	65.6	3.0	2.0	74	48.5	46.1	26	17.1	1.0	0.2	48.2	100	446.4	48.2	446.4
1943	79.5	3.0	2.4	72	57.3	54.4	28	22.3	1.0	0.2	57.0	100	503.4	57.0	503.4
1944	95.2	3.0	2.9	70	66.6	63.3	30	28.5	1.0	0.3	66.4	100	569.8	66.4	569.8
1945	87.5	3.0	2.6	68	59.5	56.5	32	28.0	1.0	0.3	59.4	100	629.2	59.4	629.2
1946	67.4	3.0	2.0	66	44.5	42.3	34	22.9	1.0	0.2	44.5	100	673.8	44.5	673.8
1947	90.4	3.0	2.7	64	57.9	55.0	36	32.5	1.0	0.3	58.0	100	731.8	58.0	731.8
1948	97.4	3.0	2.9	62	60.4	57.4	38	37.0	1.0	0.4	60.7	100	792.5	60.7	792.5
1949	89.4	3.0	2.7	60	53.6	50.9	40	35.7	1.0	0.4	54.0	100	846.4	54.0	846.4
1950	98.2	2.5	2.5	58	57.0	54.1	42	41.3	1.0	0.4	57.0	100	903.4	57.0	903.4
1951	110.9	2.5	2.8	56	62.1	59.0	44	48.8	1.0	0.5	62.3	99	965.7	62.9	966.3
1952	99.5	2.5	2.5	53	52.7	50.1	47	46.8	1.0	0.5	53.1	98	1018.7	54.1	1020.5
1953	117.8	2.5	2.9	50	58.9	56.0	50	58.9	1.0	0.6	59.5	97	1078.2	61.3	1081.8
1954	106.5	2.5	2.7	46	49.0	46.6	54	57.5	1.0	0.6	49.8	96	1128.0	51.9	1133.7
1955	130.4	2.5	3.3	42	54.8	52.0	58	75.6	1.0	0.8	56.0	95	1184.1	59.0	1192.6
1956	137.2	2.5	3.4	39	53.5	50.8	61	83.7	1.0	0.8	55.1	93	1239.2	59.3	1251.9
1957	144.2	2.5	3.6	34	49.0	46.6	66	95.1	1.0	1.0	51.1	91	1290.3	56.2	1308.1
1958	141.9	2.5	3.5	29	41.2	39.1	71	100.8	1.0	1.0	43.7	89	1333.9	49.1	1357.1
1959	166.8	2.0	3.3	27	45.0	42.8	73	121.8	1.0	1.2	47.3	87	1381.3	54.4	1411.5
1960	168.8	2.0	3.4	25	42.2	40.1	75	126.6	1.0	1.3	44.7	85	1426.0	52.6	1464.2
1961	174.1	2.0	3.5	23	40.0	38.0	77	134.0	1.0	1.3	42.9	82	1468.9	52.3	1516.4
1962	219.4	2.0	4.4	21	46.1	43.8	78	173.3	1.0	1.7	49.9	79	1518.8	63.2	1579.6
1963	235.5	2.0	4.7	19	44.7	42.5	81	190.8	1.0	1.9	49.1	77	1567.9	63.8	1643.4
1964	243.1	2.0	4.9	17	41.3	39.3	83	201.8	1.0	2.0	46.1	74	1614.0	62.3	1705.7
1965	269.3	2.0	5.4	16	43.1	40.9	84	226.2	1.0	2.3	48.6	72	1662.6	67.5	1773.2
1966	293.9	2.0	5.9	14	41.1	39.1	86	252.8	1.0	2.5	47.5	70	1710.1	67.9	1841.0
1967	323.7	2.0	6.5	13	42.1	40.0	87	281.6	1.0	2.8	49.3	68	1759.4	72.4	1913.5
1968	346.3	1.5	5.2	12	41.6	39.5	88	304.7	1.0	3.0	47.7	66	1807.1	72.3	1985.8
1969	400.4	1.5	6.0	11	44.0	41.8	89	356.3	1.0	3.6	51.4	65	1858.5	79.1	2064.9
1970	458.7	1.5	6.9	9	41.3	39.2	91	417.4	1.0	4.2	50.3	63	1908.8	79.8	2144.7
1971	457.8	1.5	6.9	8	36.6	34.8	92	421.1	1.0	4.2	45.9	61	1954.6	75.2	2219.9
1972	452.1	1.5	6.8	7	31.6	30.1	93	420.4	1.0	4.2	41.0	59	1995.7	69.6	2289.4
1973	475.0	1.5	7.1	5	23.8	22.6	95	451.3	1.0	4.5	34.2	58	2029.9	59.0	2348.4
1974	527.3	1.5	7.9	5	26.4	25.0	95	501.0	1.0	5.0	38.0	56	2067.9	67.8	2416.2
1975	413.2	1.5	6.2	5	20.7	19.6	95	392.6	1.0	3.9	29.8	55	2097.6	54.1	2470.3
1976	388.0	1.5	5.8	5	19.4	18.4	95	368.6	1.0	3.7	27.9	52	2125.5	53.6	2524.0

*Total Emission = Production Loss + Dispersive Loss + Consumptive Loss.

SOURCE: See Figure 5.2.

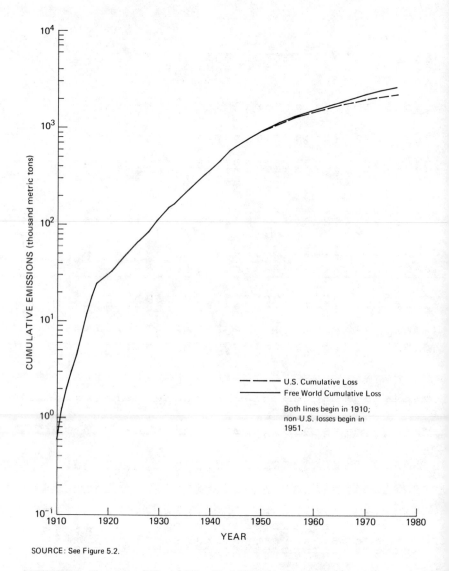

FIGURE 5.3 Cumulative U.S. and "Free World" emissions of carbon tetrachloride, 1910-1976.

TABLE 5.8 Comparisons of Findings on Annual and Cumulative Emissions of Carbon Tetrachloride

Base Year	Annual Emissions (10^3 metric tons/year)		Cumulative Emissions (10^6 metric tons)		Reference
	U.S.	"Free World"	U.S.	"Free World"	
1973	20.9[a]	41.7			A.D. Little, Inc. (1975)
1973	40.9	81.8			McCarthy (1974)
1973	41.7	96.4	1.9	2.5	Singh et al. (1976)
1974	23.0[b]	47.5	1.76	1.8	Altshuller (1976)
1973	45	90			Galbally (1976)
1973	34.2	59.0	2.03	2.35	Panel estimates
1974	38.0	67.8	2.07	2.42	"
1975	29.8	54.1	2.10	2.47	"
1976	27.9	54.2	2.13	2.52	"

[a]Uncorrected for emissions not reaching the atmosphere, to permit comparison on same basis.
[b]May not include losses during production and consumptive uses.

off or decreased in recent years as shown in Figure 5.2. The two studies that explored cumulative "free world" emissions of CCl_4 are also in fairly close agreement on those amounts. Cumulative emissions of about 2.1×10^6 metric tons for the United States and 2.5×10^6 metric tons worldwide are estimated by this Panel through 1976.

Attempts also have been made to account for atmospheric levels of CCl_4 (see Analysis of Global Mass Balances, this chapter). Galbally and others suggest that anthropogenic emissions account for a large part of all CCl_4 observed in the atmosphere. Singh et al. (1976) suggested that heterogeneous decomposition of trichloroacetyl chloride, an assumed product of the photooxidation of perchloroethylene, might be a secondary anthropogenic source of carbon tetrachloride. Assuming that about 8 percent of all perchloroethylene emitted to the troposphere may eventually be converted to CCl_4, the authors say this source could account for 0.7×10^6 metric tons of CCl_4, or about 35 percent of cumulative world emissions. However, this formation reaction has been reported only for a laboratory smog chamber, where it is presumed that chlorine radicals may be present. At present it is not known if similar formation can occur in the atmosphere. This source was not included in any calculations in the literature or in this report because further research is needed to establish unequivocally whether such reactions occur in the atmosphere.

Other Halomethanes

Historical emissions of other halomethanes have not been documented. Direct anthropogenic emissions of chloroform probably have changed over the years in a manner similar to, but not as dramatically, as those of carbon tetrachloride. Uses of chloroform before the advent of fluorocarbon aerosols in the 1950s tended to be largely dispersive. However, its production rate always has been considerably lower than that of carbon tetrachloride. A reasonable estimate would be that chloroform emissions have been about 25 percent of those of carbon tetrachloride.

Emissions of chloroform from such secondary anthropogenic sources as liquid industrial wastes and municipal water supply treatment have not been quantified. The formation of chloroform from trichloroethylene via photodegradation has been reported in smog chamber studies. The mechanism is probably similar to that of the formation of CCl_4 from perchloroethylene. Further study is needed to determine if the formation of chloroform from trichloroethylene is significant in the real atmosphere (Appleby et al. 1976b).

The uses for methylene chloride and methyl chloride have remained about the same over the years and have not led to significant variations in emission patterns. Based on cumulative U.S. production estimates for each compound through 1976 (Table 4.2) and assuming losses to the environment of 84.2 percent for methylene chloride and 2.1 percent for methyl chloride (Table 5.1), cumulative emissions for the United States can be calculated. Assuming the regional percentages shown on Table 4.3, these emissions can be extrapolated to world values. These extrapolated values do not include emissions generated by chlorination and combustion processes which have not been quantified.

Estimates of the cumulative United States and "free world" emissions from product manufacture and use of all four chloromethanes are summarized in Table 5.9. These data are based upon all preceding assumptions regarding market distribution and extent of losses in various subcategories. The principal cumulative U.S. emissions are carbon tetrachloride (41 percent), a compound with a long tropospheric lifetime, and methylene chloride (47 percent), which has a relatively short lifetime in the atmosphere.

Natural Sources

The information available on formation of halomethanes from natural sources is scarce. Sources are just beginning to be identified and quantitative estimates of contributions generally cannot be made. The greatest amount of information is available for methyl chloride, which is produced from forest fires and agricultural burning; in reactions in seawater; and possibly from marine plants. Although methyl chloride and methyl bromide have significant anthropogenic sources, these compounds are thought to be primarily natural in origin. No significant anthropogenic sources of methyl iodide are known.

Of the other halomethanes with natural sources (methylene chloride, the haloforms, carbon tetrachloride, and several mixtures), none presently are believed to contribute significantly to global concentrations.

Forest Fires and Agricultural Burning

Palmer (1976) recently estimated U.S. production of methyl chloride in combustion as a source of 2.1×10^{11} g/yr, and that 40 percent of this amount comes from burning polyvinyl chloride in waste, with the rest stemming from forest fires and agricultural burning, (1.26×10^{11} g/yr). A proportional amount would be expected to be produced from agricultural burning and forest fires worldwide.

TABLE 5.9 Estimated Cumulative Emissions of Chloromethanes (1908-1976)[a] From Product Manufacture and Use

Compound	United States		"Free World"	
	10^6 Metric Tons	Percent	10^6 Metric Tons	Percent
Carbon Tetrachloride	2.13	41.4	2.52	32.9
Chloroform	.53	10.3	.63	8.2
Methylene Chloride	2.44	47.4	4.44	57.9
Methyl Chloride	0.5	.9	.08	1.0
Total	5.15	100.0	7.67	100.0

[a] Based on production, transport, and storage losses from consumptive and dispersive uses.

Smith and Suggs (1976) report that Fritschen et al. detected the presence of methylene chloride in smoke from a broadcast burn in a U.S. forest during the summer of 1968. Gas chromatograph-mass spectrometer analysis detected methylene chloride at three of seven locations sampled, in concentrations of 0.11, 0.48, and 0.65 ppm.

Oceans and Marine Biota

Oceans generally are believed to be the major source of these monohalomethanes: methyl chloride, methyl bromide, and methyl iodide. This belief is based primarily on the fact that identified anthropogenic sources cannot account fully for measured tropospheric concentrations, and it is supported by theories of monohalomethane formation by various mechanisms that have been corroborated to different degrees. For example, methyl chloride may be the product of a reaction between methyl iodide and the chloride ion of seawater (Lovelock 1975). However, reliable estimates of quantities of the monohalomethanes formed in the oceans cannot be made with the information available.

Biological production of haloforms in red marine algae has been reported. Burreson et al. (1975) reported that the essential oil of *Asparagopsis taxiformis*, a red marine alga or seaweed eaten by Hawaiians, is composed mainly of haloforms. Bromoform ($CHBr_3$) alone comprises an estimated 80 percent by weight of the oil, with $CHBr_2I$ comprising 5 percent and $CHBrI_2$ 2 percent (Moore et al. 1976). Another species of red alga, *Asparagopsis armata*, also has been analyzed for its halocarbons (Su and Goldberg 1976). $CHBrCl_2$, CH_2ClI or $CBrCl_3$, CH_3I, $CHBr_2Cl$, $CHBr_3$, $CHCl_3$ and CCl_4 were found. Carbon tetrachloride and chloroform were detected in similar concentrations within the alga.

Moore and his colleagues have invoked the "haloform reaction" to explain in part the occurrence of the haloforms. This classic reaction involves the interactions of methyl ketones with hypochlorite, hypobromite, or hypoiodite to produce haloforms. The successive replacement of three hydrogens on the same carbon atom by halogens clearly can give rise to a variety of halocarbons.

However, the quantitative importance of algae in providing haloforms to the environment has yet to be established. First, the concentrations of the halocarbons in various algae have not yet been determined. Second, the annual production of algal material containing halocarbons must also be estimated to ascertain the importance of such reactions. The work of Moore and his colleagues has initiated work in other laboratories along these lines and more detailed data are expected soon. The contribution from

other marine plants, such as methyl chloride from kelp
(Lovelock 1975), is also yet to be determined.

Atmospheric Production

Lovelock (1974) suggested a possible mechanism for
atmospheric production of carbon tetrachloride. He noted
that the abundance of CCl_4 in the atmosphere does not differ
appreciably between the northern and southern hemispheres.
In addition, he pointed out that the occurrence of
fluorocarbons, clearly of human origin, in air arriving at
western Ireland showed a strong correlation with that air
having arrived from continental Europe; but no such
correlation is observed with carbon tetrachloride. Finally,
he carried out some preliminary laboratory experiments in
glass vessels in which he observed the production of small
but significant quantities of carbon tetrachloride from the
reactants methane and chlorine gas. These findings prompted
Lovelock to favor the theory of an atmospheric production of
carbon tetrachloride--perhaps some heterogeneous reaction,
and not a direct gas phase chlorination of methane, which is
kinetically unfavorable. He suggests that chloroform and
unsymmetrical trichloroethane may be similarly produced.

Taking Lovelock's theory into consideration, it is still
most likely, as Graedel and Allara (1976) concluded, that
chloroform, carbon tetrachloride, and other chlorinated
hydrocarbon formation from natural atmospheric reactions is
negligible. An analysis of possible thermal and
photochemical atmospheric reactions for the production of
CCl_4, $CHCl_3$, CH_3I, CH_3CCl_3 and chlorinated ethylenes
indicated that they could not account for their observed
atmospheric concentrations.

ENVIRONMENTAL CONCENTRATIONS

Marine Water

Only four of the compounds studied in this report have
been measured in seawater: methyl iodide, methyl chloride,
chloroform, and carbon tetrachloride.

Methyl chloride was measured in surface seawaters on the
southern English coast by Lovelock (1975), who found a large
but variable concentration range (5.9-21 x 10^{-9} ml of gas
per ml of water).

Four sets of analyses of either chloroform or carbon
tetrachloride or both have been performed in seawaters.
These data are shown in Table 5.10. The only distributional
pattern that emerges is that carbon tetrachloride is less

TABLE 5.10 Chloroform and Carbon Tetrachloride Concentrations in Seawaters (10^{-3} μg/l)

Location	Chloroform	Carbon Tetrachloride	Reference
Open Ocean, East Pacific (mixed layer, 0-100 meters)	14.8 ± 5.3	0.51 ± 0.28	Su and Goldberg (1976)
Scripps Institution of Oceanography pier water (1/28/75 to 7/8/75)			
Including rainy season	11.8 ± 5.8	0.67 ± 0.17	Su and Goldberg (1976)
Excluding rainy season	9.3 ± 3.6	0.72 ± 0.06	
Northeast Atlantic	8.3 ± 1.8	0.17 ± 0.04	Murray and Riley (1973)
Liverpool Bay	–	0.25[a]	Lovelock et al. (1973)
Atlantic Ocean	–	0.41	Pearson and McConnell (1975)

[a] Includes trichloroethane (CH_3CCl_3).

abundant than chloroform in seawater. Insufficient
information is available to say more.

Precipitation and Fresh Water

Some measurements of halogenated methanes in the
hydrosphere have been reported by Su and Goldberg (1976).
They note that chloroform is more abundant than carbon
tetrachloride and that, except for hot springs, the
concentrations of these and other halocarbons are similar in
precipitation and other parts of the hydrosphere. They
report typical concentrations as follows:

	Approximate Concentration (µg/l)		
	Methyl Iodide	Chloroform	Carbon Tetrachloride
Rain (La Jolla, Calif.)	1.3 ± 0.9	17 ± 13	2.8 ± 2.2
Snow (North America)	3 - 5	3 - 90	0.3 - 2
Untreated Reservoir	1.4 ± 0.6	11 ± 4	1.4 ± 0.09

In the National Organics Reconnaissance Survey (NORS)
for Halogenated Organics in Drinking Water, which was
conducted by EPA in 1975, measurements were made of the
concentrations of chloroform, bromoform, carbon
tetrachloride, bromodichloromethane, dibromochloromethane,
and 1,2-dichloroethane in both the raw water and finished
municipal waters of 80 cities in the United States (Symons
et al. 1975). Bromoform and dibromochloromethane were not
detected in any of the raw waters at sensitivity limits of
about 5 and 0.1 µg/l, respectively. Chloroform was present
in the raw waters of 49 locations at concentrations
generally less than 0.1 µg/l. Bromodichloromethane was
present in seven of the raw water samples at concentrations
ranging from <0.2-0.8 µg/l. Carbon tetrachloride was found
in only four of the raw waters sampled, with a maximum
concentration of 4 µg/l; confirmatory analyses, made by a
second technique at several locations, resulted in a maximum
concentration of less than 2 µg/l.

In the Region V Organics Survey conducted by EPA (1975),
measurements were made of the concentrations of the same
compounds and of methylene chloride in raw water supplies at
83 locations. The mean concentration of each compound in

the raw water was less than 1 µg/l (no bromoform was found). The maximum concentrations are shown below:

Compound	Percent of Locations with Positive Results	Maximum Concentration (µg/l)
Bromoform	none	---
Chloroform	27	94.0
Bromodichloromethane	5	11.0
Dibromochloromethane	2	1.4
Carbon Tetrachloride	18	20.0
Methylene Chloride	1	1.0

Bellar et al. (1974) found similar halogenated methane concentrations in raw waters. Because their study involved both raw and treated waters, these data are shown with the data on finished drinking water in the next section of this report.

Finished Drinking Water

Bellar et al. (1974) studied the incidence and possible formation of the low molecular weight halogenated hydrocarbons in municipal drinking waters and found that the highest concentrations of these compounds were in finished waters that had surface waters as their sources (Table 5.11) and that chlorinated hydrocarbons form as a result of the chlorination process during water treatment (Table 5.12). Halogenated compounds were also measured at various stages in a sewage treatment process and found to be present at varying concentrations (Table 5.13). The concentration of chloroform increased following chlorination.

In its *Preliminary Assessment of Suspected Carcinogens in Drinking Water*, EPA (1975) qualitatively identified 10 halogenated methanes in finished drinking waters in this country: methyl bromide, methyl chloride, methylene bromide, methylene chloride, bromoform, chloroform, carbon tetrachloride, bromodichloromethane, dibromochloromethane, and dichloroiodomethane.

In the NORS study, performed by EPA in 80 cities, where halogenated hydrocarbons were found in the finished waters they usually were at greater concentrations than in the raw waters (Symons et al. 1975). When detected, the halogenated methanes were present at less than 4 µg/l in the raw waters. In the finished waters, the number of positive samplings and concentrations were:

TABLE 5.11 Trihalogenated Methane Content of Various Treated Municipal Water Supplies

Code to Sampling Site	Raw Water Source of Treated Waters	Date Sample Collected	Concentration (µg/l)		
			Chloroform	Bromo-dichloro-methane	Dibromo-chloro-methane
100[a]	Surface	8-73	94.0	20.8	2.0
100[a]	Surface	2-74	37.3	9.1	1.3
101[b]	Surface	2-74	70.3	10.2	0.4[c]
102[b]	Surface	2-74	152.0	6.2	0.9[c]
103[b]	Surface	2-74	84.0	2.9	<0.1
104[a]	Well	8-73	2.9	No data	No data
104[a]	Well	2-74	4.4	1.9	0.9[c]
105[a]	Well	2-74	1.7	1.1	0.8[c]
106[b]	Well	12-73	3.5	No data	No data

[a] Sample age <4 hours.
[b] Sample age unknown, >24 hours.
[c] Approximate value ±20%.

SOURCE: Bellar et al. (1974).

TABLE 5.12 Trihalogenated Methane Content of Water at Water Treatment Plant

Sample Source	Sampling Point	Free Chlorine (ppm)	Concentration (µg/l)		
			Chloroform	Bromo-dichloro-methane	Dibromo-chloro-methane
Raw river water	1	0.0	0.9	a	a
River water treated with chlorine and alum-chlorine (contact time ~80 min).	2	6	22.1	6.3	0.7
3-day-old settled water	3	2	60.8	18.0	1.1
Water flowing from settled area to filters[b]	4	2.2	127	21.9	2.4
Filter effluent	5	Unknown	83.9	18.0	1.7
Finished water	6	1.75	94.0	20.8	2.0

[a] None detected. If present, the concentration is <0.1 µg/l.
[b] Carbon slurry added at this point.

SOURCE: Bellar et al (1974).

TABLE 5.13 Organochlorine Compounds in Water at Sewage Treatment Plant

Compound[a]	Concentration (μg/l)		
	Influent before Treatment	Effluent before Chlorination	Effluent after Chlorination
Methylene chloride	8.2	2.9	3.4
Chloroform	9.3	7.1	12.1
1, 1, 1-Trichloroethane	16.5	9.0	8.5
1, 1, 2-Trichloroethylene	40.4	8.6	9.8
1, 1, 2, 2-Tetrachloroethylene	6.2	3.9	4.2
Σ Dichlorobenzenes	10.6	5.6	6.3
Σ Trichlorobenzenes	66.9	56.7	56.9

[a] All confirmed by gas chromatography-mass spectrometry.

SOURCE: Bellar et al. (1974).

Compound	Number of Cities with Positive Results	Concentration, µg/l		
		Minimum	Median	Maximum
Chloroform	80	<0.1	21.0	311
Bromodichloromethane	78	0.3	6.0	116
Dibromochloromethane	72	<0.4	1.2	110
Bromoform	26	<0.8	(a)	92
Carbon Tetrachloride	10	<2.0	--	3

(a) 93.8% of 80 cities had ≤ 5 µg/l bromoform.

The frequency distributions for the finished waters are shown in Figure 5.4.

Symons et al. (1975) conclude from this survey that the four trihalomethanes result from chlorination and are widespread in chlorinated drinking waters across the United States. In general, the concentrations of these compounds were related to the organic content of the raw water. Higher concentrations of the compounds were found when surface water was the raw water source, when raw water chlorination was practiced, and when more than 0.4 mg/l free chlorine residual was present.

In another part of the NORS study an effort was made to develop a more comprehensive survey of the organic content of finished waters. For this purpose, an extensive analysis was done of drinking waters from five major categories of raw water sources. The following localities and their sources were selected for study:

Water Source	Localities
Ground water	Miami, Fla.
	Tucson, Ariz.
Uncontaminated upland water	Seattle, Wash.
	New York, N.Y.
Raw water contaminated with agricultural runoff	Ottumwa, Iowa
	Grand Forks, N.Dak.
Raw water contaminated with municipal waste	Philadelphia, Pa.
	Terrebonne Parish, La.
Raw water contaminated with industrial waste	Cincinnati, Ohio
	Lawrence, Mass.

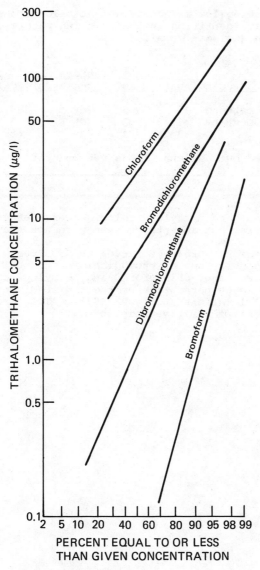

SOURCE: U.S. EPA (1975).

FIGURE 5.4 Frequency distribution of halogenated methane concentrations found in NORS study of drinking water in 80 U.S. cities.

The following low molecular weight halocarbons were quantitatively identified in the drinking waters of 5 of the 10 localities (Coleman et al. 1976):

Compound	Concentrations, µg/l (ppb)				
	Cincinnati	Miami	Ottumwa	Philadelphia	Seattle
Bromodichloromethane	15	73	(d)	20	4
Dibromochloromethane	3	32	(d)	5	3
Chloroform	38	301	1	65	21

(d) = detected but concentrations not quantified

Methyl chloride and methylene chloride were detected in the same five cities but amounts present were not quantified.

In its Region V Organics Survey at 83 sites, EPA also reported concentrations of bromodichloromethane, dibromochloromethane, chloroform, and methylene chloride in a relatively large percentage of the finished municipal waters at all locations (U.S. EPA 1975). The concentrations and the percent of positive samplings were:

Compound	Percent of Locations with Positive Results	Concentrations (µg/l) Median	Maximum
Bromodichloromethane	78	6	31
Dibromochloromethane	60	1	15
Chloroform	95	20	366
Bromoform	14	<1	7
Carbon Tetrachloride	34*	<1*	26*
Methylene Chloride	8	<1	7

* A total of eleven samples may have been contaminated by exposure to laboratory air containing carbon tetrachloride.

On the basis of the Region V Organics Survey, it appears that:

- Raw water with low turbidity (relatively low organics) results in finished water that is relatively free of chloroform and related compounds.

- The four trihalomethanes--bromodichloromethane, dibromochloromethane, chloroform, and bromoform--result from the chlorination of organic precursor materials in raw water. The other compounds--carbon tetrachloride and methylene chloride--do not seem to be formed during chlorination.

- There are correlations among the concentrations of the four trihalomethanes that seem to point to common precursors.

- The use of powdered activated carbon was inefficient in removing the volatile organic compounds.

Morris and McKay (1975) critically reviewed the status of knowledge concerning the incidence and formation of halogenated organic compounds during the chlorination of water supplies. He concluded that chloroform or other haloforms can result as end products in water chlorination but that general substitution reactions of chlorine are unlikely. Thus, carbon tetrachloride or fully chlorinated higher hydrocarbons probably are not products of water chlorination. He also concluded that there are available methods for minimizing the concentrations of these compounds in water supplies, including pre-treatment methods, such as coagulation or pre-ozonation to reduce the amounts of precursor compounds, and post-treatment methods, such as carbon adsorption or aeration to remove them after halomethane formation.

McKinney et al. (1976) conducted studies for several months in 1975 on the volatile organic levels of tap water from a municipal supply system in a laboratory of the National Institute of Environmental Health Sciences (NIEHS). They found that the monthly average chloroform concentration in the tap water varied seasonally, with a low of 407 µg/l in January and a high of 1,384 µg/l in August. There was also considerable daily variation. These are their data for average monthly chloroform measurements (in µg/l):

January	407.2	June	650.1
February	425.3	July	1,064.1
March	429.6	August	1,384.0
April	562.2	September	1,125.7
May	565.1	October	1,112.3

The authors say values may be high by about a factor of two. Even when corrected the upper limits appear to be somewhat higher than those found in the NORS and Region V studies.

McKinney et al. also sought to correlate the chloroform levels with various factors. Figure 5.5 shows a plot of the weekly average chloroform concentrations with the temperature and chlorine feed. The chloroform level varies reasonably well with the temperature data, with lows occuring in January and February and highs in August, and also with the chlorine feed. The chlorine feed, while varying by a factor of three, is adjusted to maintain a fairly constant 2.0 to 2.5 ppm chlorine residual and reflects changes in the total oxidizable dissolved organics content and in the rates of various oxidation reactions, including chlorination. It appears that the higher temperatures and the increased chlorine feed concentration in summer increase the amount of chlorination of the organics to form chloroform. Although there is a higher total organic content in the winter, more extensive degradation of the volatile organics by chlorination occurs in the summertime.

Indoor Atmosphere

Recently extensive measurements of halomethanes concentrations in various indoor atmospheres have been made by Harsch (1977) and Harsch and Rasmussen (1977). The

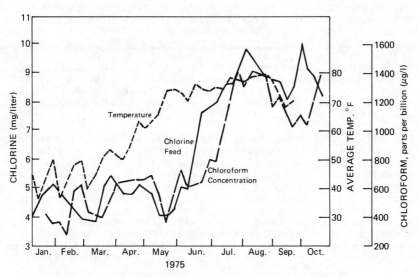

SOURCE: McKinney et al. (1976) Possible factors in the drinking water of laboratory animals causing reproductive failure. In *Identification and Analysis of Organic Pollutants in Water*, Lawrence H. Keith, editor. Ann Arbor Science Publishers, Inc.

FIGURE 5.5 Correlation of temperature and chlorine feed with chloroform concentrations in tap water samples.

results for methyl chloride, methylene chloride, chloroform, and carbon tetrachloride are shown in Table 5.14, those for methyl bromide are shown in Table 5.15.

The concentration of methyl chloride ranged from low values of 525 parts per trillion (ppt) (v/v) to a maximum concentration of 20,000+ ppt (v/v). The lowest concentrations (500-700 ppt (v/v)) reflect the background levels generally found in continental tropospheric samples. It is interesting that most of the samples showed methyl chloride levels considerably elevated over background, although there are not many direct anthropogenic sources of the compound. One explanation for this may be provided by the concentration of over 20 ppb (v/v) found in an apartment after a cigarette had been smoked (Table 5.14). Combustion produces considerable amounts of methyl chloride (Palmer 1976), and burning tobacco is a significant localized source of methyl chloride (see Chapters 5 and 6). Generally, the indoor concentration of methyl chloride is between two and ten times the ambient background level, and in nearly all cases it is the predominant halomethane in the indoor atmosphere.

In the few cases where methyl chloride was not the major halomethane, methylene chloride (CH_2Cl_2) was. In most cases, a local source of methylene chloride is obvious. For example, a concentration of 23.4 ppb (v/v) found in a beauty parlor was attributable to extensive use of aerosol hair sprays, and a concentration of 4.2 ppb (v/v) was reported from the solvent section of a carpet store. Virtually all of the indoor air samples have methylene chloride concentrations above the tropospheric background level of 20 to 50 ppt (v/v). Indoor air has between 10 and 1,000 times the amount of CH_2Cl_2 found in unpolluted tropospheric air.

The concentration of chloroform in indoor air, on the other hand, was rarely over 100 ppt (v/v), and never approached the methyl chloride or methylene chloride levels except in the one sample of car exhaust which had 5.6 ppb (v/v). Most of the indoor air chloroform measurements are very close to its tropospheric background concentration of 20 to 40 ppt (v/v), and even the maximum concentrations found in indoor air are lower than the values reported for outside air in many urban areas.

Like that of chloroform, the concentration of carbon tetrachloride in indoor air is usually fairly low, between 120 and 300 ppt (v/v). The lowest concentration measured, 19 ppt (v/v) in the exhaust of a car with a catalytic converter, is not representative, but most of the concentrations in the range from 120 to 150 ppt (v/v) probably represent background tropospheric concentrations over North America and throughout the Northern Hemisphere.

TABLE 5.14 Chlorinated Methane Concentrations in Indoor Atmospheres.

Sample Source	Concentration (ppt, by volume)			
	Methyl Chloride	Methylene Chloride	Chloroform	Carbon Tetrachloride
Discount Store				
Records & automotive	4,000+	14,000+	25	386
Plastic wares	1,500+	NA	19	238
Shoes & furniture	1,500+	NA	22	256
Restaurant				
Tables & booth	1,450+	380	14	143
Counter	NA	NA	36	143
Kitchen	NA	NA	150	144
Movie Theater				
Lobby before show	900	850	37	155
Theater before show	700	890	29	156
Lobby after show	1,130	525	81	196
Theater after show	2,500+	525	78	191
Body Shop				
Paint storage	525	570	27	154
Painting room	NA	NA	16	162
Offices & Classroom				
Xerox room	760	170	104	238
Typing room	905	160	35	211
Classroom	650	185	190	326
Air Terminal				
Baggage & ticket area	1,240	190	29	152
Restaurant & waiting	1,405	730	55	153
Concourse	1,400	135	78	152
Boeing 727 Airplane				
First class	750	340	47	210
Coach, front	1,260	160	24	125
Coach, rear	1,000	110	38	122
Cocktail Lounge				
Bar	900	8,000	213	177
Tables	1,000+	NA	248	165
Beauty Parlor				
Waiting area	1,320	23,400+	110	246
Dryers	NA	NA	115	273
Hair Styling				
Waiting area	NA	NA	283	149
Chairs	NA	NA	496	159
Floor Covering Store				
Carpet storage	1,000+	NA	17	305
Solvent shelves	1,100	4,210	22	459
Tile area	2,000+	NA	23	357

TABLE 5.14 (Continued)

	Concentration (ppt, by volume)			
Sample Source	Methyl Chloride	Methylene Chloride	Chloroform	Carbon Tetrachloride
Automobile				
New 1976 Ford, motor off	2,000+	NA	41	229
1972 Rambler, motor off	NA	NA	29	155
1972 Rambler with heater on, driving (leaded fuel)	3,000+	NA	35	147
1975 Pinto, motor off	650	180	32	147
1973 Dodge, motor off	8,000	64	20	144
1973 Dodge with air conditioner, running (leaded fuel)	NA	NA	23	144
1975 Pinto exhaust (non-leaded fuel, catalytic converter)	700	140	33	19
1972 Rambler exhaust (leaded fuel)	1,000+	91	5,600	131
Drug & Variety Store				
Cards & candles area	NA	NA	337	175
Photo & drug area	NA	NA	473	172
Cosmetics area	1,100	2,900	265	165
Urban Street Corner				
AM early	NA	NA	8	131
light traffic	1,000+	NA	20	149
Rush hour	750	670	14	147
Radio & TV Repair				
Repair area	4,000+	300	37	184
Counter	NA	NA	44	201
Display floor	NA	NA	41	209
Automobile Dealership				
Shop	NA	NA	15	133
Display floor	3,000+	7,160+	33	150
Food Store				
Produce area	1,200	6,100+	647	231
Cleaners	NA	NA	733	255
Frozen food	NA	NA	641	231
Laundromat				
Washers	570	60	67	150
Dryers	1,000+	NA	49	146
Dry cleaners	NA	NA	49	140

TABLE 5.14 (Continued)

Sample Source	Concentration (ppt, by volume)			
	Methyl Chloride	Methylene Chloride	Chloroform	Carbon Tetrachloride
Apartment				
While cleaning				
bathroom	700	800	134	178
PM	525	200	49	163
AM	NA	NA	43	161
After shower	NA	NA	61	280
Party	2,500+	390	46	282
After a cigarette smoked	20,000+	420	59	146
Drive-in Restaurant				
Dining area, inside	NA	NA	300	172
Food preparation area	NA	NA	297	172
Order window, inside	1,790	5,740+	271	182

NA = Not available.

SOURCE: Harsch (1977).

TABLE 5.15 Concentrations of Methyl Bromide in Urban Air, Automobile Exhaust

Date	Sample Description	Methyl bromide (ppt)
9/13/76	Street corner, moderate traffic, 10 mph wind	<10
9/13/76	Same, heavy traffic	<10
10/1/76	Automobile exhaust, leaded fuel, 1972 Rambler	55,000
10/1/76	Auto exhaust, non-leaded fuel, catalytic convertor 1975 Ford Pinto	<1,000
11/6/76	Street corner, light traffic, calm early morning with heavy inversion	220
11/6/76	Same, heavy traffic, slight breeze	150
11/16/76	Street corner, light traffic, 5 mph wind	<10
11/16/76	Same, heavy traffic	185
11/16/76	Auto exhaust, leaded fuel, 1972 Rambler	18,000
11/16/76	Auto exhaust, non-leaded fuel, catalytic convertor 1975 Ford Pinto	1,300

SOURCE: Harsch and Rasmussen (1977).

Again, like chloroform, the maximum CCl_4 concentrations in indoor air are considerably lower (up to 40 times less) than some values reported for outside air in urban areas.

Finally, Harsch and Rasmussen (1977) have recently made measurements of methyl bromide in automobile exhaust. While exhaust is not really an indoor atmosphere, elevated concentrations could occur in a closed garage, although carbon monoxide is a more serious danger in that case. Their measurements show that a car burning leaded gasoline had from 18 to 55 ppb (v/v) methyl bromide in its exhaust, while a car equipped with a catalytic converter burning unleaded gasoline had only 1 to 2 ppb (v/v) methyl bromide in its exhaust. Methyl bromide is a breakdown product of ethylene dibromide (CH_2Br-CH_2Br), which is a gasoline additive. They also found that concentrations as high as 220 ppt (v/v) of methyl bromide could occur outdoors, with light traffic, during an early morning inversion.

In sum, methyl chloride and methylene chloride seem to occur in relatively higher concentrations in indoor air than in the outside atmosphere, sometimes orders of magnitude higher where there are local sources. On the other hand, chloroform and carbon tetrachloride in indoor atmospheres generally occur in concentrations which reflect their outside atmospheric levels, and are actually lower than the maximum concentrations found in urban air. Methyl bromide occurs in elevated concentrations in the exhaust of cars burning leaded gasoline.

Outdoor Atmosphere

Only six of the nonfluorinated halomethanes have been measured in the atmosphere. Their continental and marine background concentrations and the range of concentrations found in urban areas are shown in Table 5.16. These values represent the best measurements available for these compounds at the present time; the background concentrations are the result of hundreds, sometimes thousands of individual measurements. Methyl chloride, chloroform, and carbon tetrachloride have been measured extensively and there is a fairly good understanding of their atmospheric distribution. However, there are relatively few measurements available for methylene chloride, methyl bromide, and methyl iodide, and there is still a great deal of uncertainty regarding their atmospheric distribution patterns.

Methyl chloride has a continental background concentration of 500 to 700 ppt (v/v), while the marine background level is approximately 1,000 to 1,300 ppt (v/v), which suggests that the oceans are a major source of the

TABLE 5.16 Summary of Atmospheric Concentrations of Nonfluorinated Halomethanes (ppt)*

Compound	Continental Background		Marine Background	Urban Range
Methyl Chloride (CH_3Cl)	569 ± 42[a] 587 ± 102[c] 713 ± 51[d] 1,040 ± 399[d] 690 ± 390[f]	530 ± 30[k]	1,260 ± 434[d] 1,140 ± 400[l]	834 ± 40[d]
Methylene Chloride (CH_2Cl_2)	36 ± 11[c]		35 ± 19[f]	<20 – 144[c]
Chloroform ($CHCl_3$)	9 ± 3[a] 11 ± 5[c] 25 ± 18[d] 20 ± 10[k]		40 ± 38[d] 27 ± 8[f]	102 ± 102[d] 10 – 15,000[g] 6 – 3,000[i]
Carbon Tetrachloride (CCl_4)	122 ± 13[a] 133 ± 10[c] 116 ± 8[d] 120 ± 15[k]		128 ± 4[b] 128 ± 16[d] 111 ± 11[f]	134 ± 20[d] 120 – 18,000[g] 1,400[h] 120 – 1,500[i]
Methyl Bromide (CH_3Br)	15 ± 10[d] 0.5 – 1.0[e]		93 ± 100[d]	108 ± 138[d] <10 – 220[e]
Methyl Iodide (CH_3I)	9 ± 5[d]		7 ± 7[d] <1 – 11[i]	24 ± 20[d] <1 – 3,800[g]

*Only those nonfluorinated halomethanes which have been measured are included on the table. The ± indicates typically one standard deviation.

[a] Cronn et al. (1976).
[b] Pierotti et al. (1976).
[c] Pierotti and Rasmussen (1976).
[d] Singh et al. (1977).
[e] Harsch and Rasmussen (1977).
[f] Cox et al. (1976).
[g] Lillian et al. (1975).
[h] Ohta et al. (1976).
[i] Su and Goldberg (1975).
[j] Lovelock et al. (1973).
[k] Grimsrud and Rasmussen (1975).
[l] Lovelock (1975).

compound. This is also supported by the fact that the concentration of methyl chloride found in Los Angeles by Singh et al. is lower than the marine background level he measured at Point Reyes, on the Pacific coast north of San Francisco (Singh et al. 1977). Mid-continental values tend to be the lowest, between 500 and 600 ppt (v/v). Although the previous section shows that relatively high concentrations of methyl chloride may occur indoors due to localized sources, direct anthropogenic sources do not seem to be significant in raising the general level of CH_3Cl in urban areas, and the highest values have been measured in the marine environment.

Data on methylene chloride in the atmosphere are extremely scarce, but recent measurements indicate that the background atmospheric concentration is about 35 ppt (v/v), with little variation between continental and marine air. Urban concentrations vary from low background levels of 20 ppt (v/v) to the maximum concentration of 144 ppt (v/v) measured near St. Louis. Like methyl chloride, methylene chloride does not seem to occur in concentrations significantly higher in urban areas than typical ambient outdoor air concentrations, although some indoor air samples showed extremely high concentrations from nearby sources. The data are insufficient at present to make any definite conclusions.

Chloroform has been measured extensively, although there still seems to be considerable uncertainty about its background concentration. Chloroform levels vary widely in the troposphere. The global background level is estimated in this study to be in the range 20 to 40 ppt (v/v), with higher concentrations occurring in marine air masses and lower values in continental air. Mid-continental concentrations of chloroform tend to be around 10 ppt (v/v). While this distribution pattern suggests a marine source of chloroform, the elevated concentrations measured in urban areas clearly point to anthropogenic sources also. The previous section on indoor atmospheres showed extremely high chloroform concentration in one car exhaust sample; this may be one reason why Singh et al. (1977) observed elevated concentrations of chloroform in Los Angeles and Palm Springs. Lillian et al. (1975) also measured extremely high chloroform concentrations in Wilmington, Ohio (4,800 ppt (v/v)) and Bayonne, New Jersey (15,000 ppt (v/v)), while Su and Goldberg (1976) found extremely high concentrations in many urban areas around the world, including Washington, D.C. (2,600 ppt (v/v)), the Chicago Loop (1,800 ppt (v/v)), downtown Brussels (3,000 ppt (v/v)), and Geneva (2,600 ppt (v/v)). Since the indoor air measurements never approached these levels, it appears that the compound is not originating from domestic or commercial uses, as in the case of the fluorocarbons, but is the byproduct of industrial processes such as paper pulp bleaching, and perhaps of combustion (including gasoline combustion), and secondary formation from polychlorinated organic compounds such as trichloroethylene.

Carbon tetrachloride is the most extensively studied of all nonfluorinated halomethanes. There have been thousands of measurements of CCl_4 made over land and the oceans, and in both the Northern and Southern Hemispheres. There are no major gradients in the atmospheric distribution of carbon tetrachloride, the concentration is very nearly the same in continental and marine air masses, from 120 to 140 ppt (v/v). The CCl_4 concentration is slightly lower in the Southern Hemisphere than in the Northern, but the gradient is decreasing as emissions in the Northern Hemisphere have stabilized, and the atmospheric distribution is approaching homogeneity. The concentration of carbon tetrachloride in urban air is generally quite close to the background level of 120 to 140 ppt (v/v), although some extremely high concentrations have been recorded. Ohta et al. (1976) obtained an annual average of 1,400 ppt (v/v) in Tokyo between April 1974 and April 1975, by far the highest level ever observed over an extended period of time. However, their values are not confirmed, and should not be taken as representative of the average concentration in an urban area. The highest CCl_4 concentrations measured in the atmosphere are Su and Goldberg's value of 1,500 ppt (v/v) for Grenoble, and the measurement of Lillian et al. (1975) of 18,000 ppt (v/v) in the extremely polluted air of Bayonne, New Jersey. Although there is considerable evidence that CCl_4 is primarily anthropogenic in origin (see Analysis of Global Mass Balances, this chapter), it has been released for such a long time that it is now the most evenly distributed of all the halocarbons in the atmosphere, with almost no variation between land and ocean, urban and rural areas, or Northern and Southern Hemispheres.

There are only a few measurements of methyl bromide in the atmosphere, but they seem to provide a relatively clear picture of its atmospheric distribution. It has a continental background concentration 20 ppt (v/v) or less, with much larger amounts in marine air and urban areas. Lovelock (1975) found high concentrations of methyl bromide in surface sea water, and concluded that the oceans are the major source of the compound. This is borne out by the high concentrations found by Singh et al. (1977) at Point Reyes. However, Harsch and Rasmussen (1977) have recently obtained clear evidence that combustion of gasoline containing the additive ethylene dibromide (CH_2Br-CH_2Br) is a significant source of the compound. This is corroborated by elevated concentrations of methyl bromide found by Singh et al. (1977) in Los Angeles, even higher than those found in marine air. Therefore, methyl bromide should be highest in urban and coastal areas, with the lowest values (less than 1 ppt (v/v)) found in rural, continental air.

Measurements of methyl iodide are almost as scarce as those of methylene chloride and there is no definite pattern to its atmospheric distribution. Singh et al. (1977) calculated a continental background of 9 ppt ±5, a marine background of 7 ppt ± 7, and an urban level of about 24 ppt ± 20. Lillian et al. (1975) found higher concentrations near the ocean than inland, in line with Lovelock's measurement of high concentrations in surface sea water. However, the higher concentrations which they associated with the ocean may simply have been from urban areas, since Singh et al. (1977) found higher methyl iodide concentrations in Los Angeles than at Point Reyes, and the highest value ever measured, 3,800 ppt (v/v), was obtained by Lillian et al. (1975) in Bayonne, New Jersey. Although there are no known anthropogenic sources of iodinated hydrocarbons, there is unequivocal evidence of some urban source of methyl iodide. Methyl iodide appears to have a distribution somewhat similar to that of methyl bromide, with the highest concentrations in urban and coastal areas, and lower values in rural areas inland.

Note should be made of the measurements made by Lillian et al. (1975) in Bayonne, New Jersey. Not only were the concentrations of methyl iodide, chloroform, and carbon tetrachloride the highest measured anywhere, but they also found the highest concentrations of the fluorocarbons and chlorinated ethylenes that have ever been measured in the atmosphere. This would seem to indicate that there may be large sources of these compounds in the Bayonne area, and that residents of the area might be subjected to much greater levels of these compounds than would the general populace. There are undoubtedly other areas of the United States where such local contamination occurs, and therefore specific communities may be exposed to concentrations 10 to 100 times as great as the rest of the country. This could have a significant impact on the human uptake in some parts of the country, particularly in the vicinity of chemical plants manufacturing these compounds.

Food

McConnell et al. (1975) reviewed the incidence, significance, and movement of chlorinated hydrocarbons in the food chain. They concluded that chloroform and carbon tetrachloride are distributed widely in the environment and are present in the air and in fresh water, rainwater, seawater, fish, water birds, marine mammals, and various foods. In foods, the typical range of chloroform and carbon tetrachloride concentrations seems to be about 1 to 30 ppb (μg/kg). Examples of these and other hydrocarbon concentrations in foodstuffs summarized by McConnell et al. are shown in Table 5.17.

McConnell et al. note that in amost all cases there are enormous differences between permitted use concentrations and the maximum levels found in foods, as shown in Table 5.18. However, they also point out that the acute toxicities of these compounds are low. The LD_{50} values for rats and mice are in the range of 1 to 10 gm/kg. Thus, by extrapolation the amounts found in food and water will not give rise to acute poisoning.

Finally, McConnell et al. note that there is no evidence for significant bioaccumulation of these compounds, via the food chain, to higher trophic levels.

Pearson and McConnell (1975) also reviewed the incidence of chlorinated hydrocarbons in various marine organisms and water birds and found that the concentrations of carbon tetrachloride and chloroform in edible fish and marine organisms ranged from 0.1 to 6 ppb (μg/kg) and 3 to 180 ppb (μg/kg), respectively.

Fumigant Residues in Foods

Fumigant residues of halogenated methanes in foodstuffs treated with fumigants containing these compounds form another source of human exposure to these compounds. Carbon tetrachloride, methyl bromide, and to a lesser extent chloroform are used as fumigants. In general, normal dosages and proper storage, handling, and aeration following fumigation will reduce the amount of the fumigant residues to low levels before commodities reach consumers.

Carbon Tetrachloride. In many countries carbon tetrachloride is used as a fumigant either alone or, more commonly, mixed with other more toxic ingredients to help distribute more toxic elements or to reduce fire and explosion hazards. Some of the commercial mixtures (they are referred to by number in subsequent text) and their percent composition, by volume are:

TABLE 5.17 Examples of Chlorinated Hydrocarbon Concentrations in Foods ($\mu g/kg$)[a]

Foodstuff	Chloroform	Carbon Tetrachloride	Trichloro-ethane	Trichloro-ethylene	Perchloro-ethylene	Others[b]
Dairy Produce						
Fresh milk	5	0.2		0.3	0.3	HCBD 0.08 HCB 1
Cheshire cheese	33	5		3	2	ND
English butter	22	14		10	13	HCBD 2
Hens' eggs	1.4	0.5		0.6	ND	ND
Meat						
English beef (steak)	4	7	3	16	0.9	ND
English beef (fat)	3	8	6	12	1.0	ND
Pigs' liver	1	9	4	22	5	TCE 0.5 PCE 0.4
Oils and Fats						
Margarine	3	6	NA	6	7	PCE 0.8
Olive oil (Spanish)	10	18	10	9	7	ND
Cod liver oil	6	16	5	19	2	TCE 2
Vegetable cooking oil	2	0.7	NA	7	0.01	HCBD 0.2 HCB 0.7
Castor oil	NA	NA	6	ND	3	NA
Beverages						
Canned fruit drink	2	0.5	NA	5	2	PCE 0.8
Light ale	0.4	0.2	NA	0.7	ND	HCBD 0.2
Canned orange juice	9	6	NA	ND	ND	ND
Instant coffee	2	5	NA	4	3	ND
Tea (packet)	18	4	7	60	3	ND
Wine (Yugoslav)	NA	0.3	NA	0.02	ND	NA
Fruits and Vegetables						
Potatoes (S. Wales)	18	8	4	ND	ND	NL
Potatoes (N.W. England)	4	3	1	3	0.7	TCE 0.7
Apples	5	5	3	5	2	ND
Pears	2	4	2	4	2	ND
Tomatoes[c]	2	4.5	NA	1.7	1.2	HCBD 0.8 HCB>0.1 TCE 1.0
Black grapes (imported)	ND	19.7	NA	2.9	ND	HCBD 3.7
Fresh Bread	2	5	2	7	1	ND

[a] NA = No Analysis; ND = Not Detected.
[b] HCBD is still used in some countries as an insecticide for vineyards. TCE, PCE = Tetra-, pentachloroethane; HCBD = Hexachlorobutadiene; HCB = Hexachlorobenzene.
[c] Tomato plants were grown on a reclaimed lagoon.

SOURCE: McConnell et al. (1975) Chlorinated hydrocarbons in the environment. *Endeavor* (Jan.):13-18.

TABLE 5.18 Maximum Concentrations of Chlorinated Hydrocarbons in Food Compared to Permitted Use Concentrations

Chemical	Uses	Permitted Concentrations/Dose	Maximum Measured Concentration (μg/kg)
Trichloroethylene	Anaesthetic, extraction solvent	10-25 mg/kg decaffeinated coffee[a]	60 (packet coffee)
Perchloroethylene	Antihelminthic (veterinary and human medicine)	1.6-8.0 g/60 kg[b] (therapeutic dose)	13 (butter)
1,1,1-Trichloroethane	?	?	20 (black grapes)
Carbon tetrachloride	Grain fumigant	50 μg/kg (cooked cereal products)[c]	20 (black grapes)
Chloroform	Anaesthetic, flavouring agent	15-30 mg/dose[b] (cough linctus)	33 (cheese)

[a] U.S. Food and Drug Administration (1973) *Food Chemical News Guide.*
[b] Martindale (1972) *The Extra Pharmacopeia,* 25th ed.
[c] FAO/WHO Expert Committees (1972) *WHO Pesticide Residues Series 1, 1971.*

SOURCE: McConnell et al. (1975) Chlorinated hydrocarbons in the environment. *Endeavor* (Jan): 13-18.

#1	Ethylene dichloride (EDC)	75%
	Carbon tetrachloride (CCl_4)	25%
#2	Carbon tetrachloride (CCl_4)	80%
	Carbon disulfide (CS_2)	20%
#3	Carbon tetrachloride (CCl_4)	60%
	Ethylene dichloride (EDC)	35%
	Ethylene dibromide (EDB)	5%
#4	Trichloroethylene ($ClCH=CCl_2$)	64%
	Carbon disulfide (CS_2)	26%
	Carbon tetrachloride (CCl_4)	10%
#5	Chloroform ($CHCl_3$)	37%
	Trichloroethylene ($ClCH=CCl_2$)	32%
	Carbon disulfide (CS_2)	26%
	Carbon tetrachloride (CCl_4)	5%

A number of foods in commerce and consumer products have been analyzed for their carbon tetrachloride residues. From 1964 to 1966, Wit (1972) analyzed a number of samples from cereals imported into the Netherlands. Carbon tetrachloride residues ranged from 0.1 to 0.5 ppm (mg/kg) in 20 percent of the samples, 0.5 to 1.0 ppm in 5 percent of the samples, 1.0 to 5.0 ppm in 8 percent of the samples, and greater than 5.0 ppm in 3 percent of the samples. The maximum CCl_4 residue found was 58 ppm. McMahon (1971) reports finding CCl_4 residues ranging from 2.9 to 20.4 ppm in commercially fumigated wheat, corn, and milo after one to three months of storage.

Bondi and Alumot (1972) analyzed commercially fumigated samples of flour and found CCl_4 residues of 0.2 to 0.3 ppm. Bread made from this flour was free of detectable carbon tetrachloride (<0.005 ppm); biscuits had a residue of 0.004 ppm.

Carbon tetrachloride residues and their behavior over time have been analyzed in laboratory experiments that simulated commercial fumigation conditions. Results of these analyses yielded residue levels similar to those in commercially fumigated commodities. Wit et al. (1972) analyzed 75 kg sacks of wheat that were fumigated with a mixture ratio of CCl_4-EDC-EDB (10.2:8:1 by weight) and then aerated for several weeks. Carbon tetrachloride residues within the sacks ranged from 20 to 62 ppm. White flour processed from this wheat had residues ranging from 2 to 10 ppm and bread from this wheat had residues up to 0.007 ppm.

Scudamore and Heuser (1973) analyzed wheat and corn samples following the application of 80 mg/l of carbon tetrachloride for three to six days. The initial residues ranged from 200 to 400 ppm; after six months of aeration, the residues ranged from 1 to 10 ppm. Residues up to 4.7 ppm were found in whole kernel wheat after 12 months aeration, indicating that carbon tetrachloride residues can be very persistent.

Bielorai and Alumot (1966) analyzed wheat and barley after treating them with 40 gm/m^3 of commercial mixture #5 for 72 hours. They found residues of 1.53 and 2.2 ppm in wheat and barley respectively on day zero, decreasing to 0.7 and 0.6 ppm respectivley by day 17. In 1969, Alumot and Bielorai (1969) analyzed several grains fumigated with commercial mixture #5 and aired at different temperatures. These data were further analyzed in 1975 (Bielorai and Alumot 1975) together with new data on carbon tetrachloride and other fumigants from both laboratory and field studies. They found more rapid desorption of fumigant residues from whole cereal grain aired at low (15°C) rather than high (30°C) temperatures, although CCl_4 was less temperature dependent than the other fumigants. However, for ground grains the temperature effect was annulled, indicating that grain structure is the principle factor involved. Bielori and Alumot suggest that unchanged fumigant residues are present in two forms, loosely and firmly bound, and that the loosely bound desorbs rapidly while the firmly bound desorbs more slowly and is the temperature dependent component.

Lynn and Vorches (1957) reported carbon tetrachloride residues in fumigated wheat and wheat fractions that were analyzed before and after treatment with commercial fumigant mixtures #2 and #3 at dosages recommended by the U.S. Department of Agriculture. The normal dosage for each mixture is 2 gallons/1,000 bushels. These residues were found:

CCl_4 Residues, parts per million (mg/kg)

Product	Before Fumigation	Normal Dosage #2 Commercial Mixture (CCl_4-CS_2)	Normal Dosage #3 Commercial Mixture (CCl_4-EDC-EDB)
Wheat (soft)	<0.5	115	(76 ppm org. Cl)
Flour	<0.5	21	10
Shorts	--	39	28
Bran	--	88	43

Product	Before Fumigation	Triple Dosage #2 Commercial Mixture CCl_4-CS_2	Triple Dosage #3 Commercial Mixture CCl_4-EDC-EDB
Wheat (soft)	<0.5	270	140
Flour	<0.5	74	34
Shorts	--	79	72
Bran	--	67	204

Berck (1974) analyzed wheat, wheat fractions, and bread fractions for carbon tetrachloride residue following treatment with Dowfume EB-5 (CCl_4, EDC, EDB; 63:30:7 percent, by weight) at a dosage of 4 imperial gallons/1,000 bushels. Samples were analyzed after different periods of aeration. The highest residues were found in the fumigated wheat; they ranged from 72.6 ppm (one week aeration) to 3.2 ppm (seven weeks aeration). The wheat fractions had residues ranging from 0.20 to 0.93 ppm for flour, 0.43 to 3.53 ppm for bran, and 0.20 to 1.65 for middlings. Bread made from wheat aerated for three days had residues of about 0.04 ppm in the upper and lower crusts, and 0.13 ppm in the crumbs, whereas in bread made from wheat aerated for seven weeks the upper crusts had no residue, the lower crust had 0.2 ppm and the crumbs had 0.01 ppm.

Results of these various studies indicate that the amount of carbon tetrachloride residue depends on the fumigant dosage, storage conditions, length of aeration, and extent of processing. Usually, proper storage and aeration reduce CCl_4 residues to trace amounts. However, several studies indicate that, despite prolonged aeration and proper storage conditions, the residues may persist at low levels for as much as a year.

Methyl Bromide. Methyl bromide is a widely used fumigant, both alone and in mixtures, in mills, warehouses, boats, railway boxcars, and nurseries. It penetrates

rapidly and deeply but its speed and efficiency are
influenced by temperature, moisture content, and the nature
of the product. Some of the common commercial fumigant
mixtures containing methyl bromide (they are referred to by
number in subsequent text) and their percent composition, by
volume, are:

#1 Methyl bromide (CH_3Br)   75%
    Ethylene dibromide (EDB)   25%

#2 Methyl bromide (CH_3Br)   36%
    Ethylene dibromide (EDB)   64%

Methyl bromide residues decrease rapidly by loss to the
atmosphere and by undergoing methylation reactions with the
protein fractions of foods, which decompose methyl bromide
and form inorganic bromide residues. Nearly all available
data on commercially fumigated foods relate to the inorganic
bromide residues in foods in commerce and consumer products
rather than to methyl bromide residues. Most investigators
agree that, due to the methylation reactions and volatility
of methyl bromide, most residual methyl bromide will quickly
disappear with proper aeration and product processing.

Laboratory fumigations have shown that proper aeration
reduces the methyl bromide residue to less than 1 ppm in a
few days. The risk that harmful amounts of free methyl
bromide will reach the consumer is insignificant. Scudamore
and Heuser (1970) investigated the residues and the
disappearance of methyl bromide in methyl bromide-fumigated
wheat, flour, cocoa beans, sultana raisins, corn, sorghum,
cottonseed meal, rice, peanuts (groundnuts), and peanut meal
under controlled conditions from both sealed and well-
ventilated samples. In all except peanuts and cocoa beans,
residual methyl bromide was reduced to less than 1 ppm
within a few days. The rate of disappearance was lower at
low temperatures and small amounts of free methyl bromide
were still being extracted from several commodities a month
after fumigation. The amount of initial methyl bromide
residues is closely proportional to the fumigant
concentration used in treatment.

Roehm et al. (1943) analyzed two samples of Longhorn
cheese fumigated with methyl bromide for total, inorganic,
and organic methyl bromide residues after various aeration
periods. Fumigation of cheese is a special use, done at low
temperatures. Bielorai and Alumot (1975) report that in
grain, fumigation temperatures affect the relative amounts
of the residues; at lower temperatures lower levels of

firmly bound residues are present. Roehm et al. (1943) found these residues in the outer 1/4 inch of the cheeses:

CH₃Br Residues, parts per million (mg/kg)

Hours of Ventilation	Longhorn Cheese A			Longhorn Cheese B		
	Inorganic	Organic	Total	Inorganic	Organic	Total
0.5	15	62	77	23	78	101
4	21	40	61	30	54	84
24	22	20	42	38	9	47
48	25	0	25	39	4	43
96	24	0	24	38	1	39
168	25	1	26	36	2	38

Seo et al. (1970) report that no methyl bromide was found in asparagus spears, avocadoes, peppers, and tomatoes after fumigation for two hours with methyl bromide at 32 mg/l.

Dennis et al. (1972) report that only trace amounts (unspecified) of residual methyl bromide was present in hard wheat flour fumigated with 36.6 mg/l methyl bromide after 9 days of aeration. The analysis of various other products had similar results.

Recommended residue levels in foods fumigated with methyl bromide usually are based on inorganic bromide residues. Table 5.19 summarizes measurements of these residues in a number of foods.

Chloroform. Chloroform is not used as extensively as carbon tetrachloride in fumigant mixtures. Its anesthetic and toxic properties, however, are useful in improving the effectiveness of insecticidal fumigant mixtures (Berck 1975). The most common commercial mixtures containing chloroform, in percent composition by weight, are:

	Mixture 1	Mixture 2
Chloroform ($CHCl_3$)	37%	38.5%
Trichloroethylene ($ClH=CCl_2$)	32%	33%
Carbon disulfide (CS_2)	26%	23%
Carbon tetrachloride (CCl_4)	5%	5.5%

Chloroform residues have not been measured in commercially fumigated commodities, but several experimental

TABLE 5.19 Maximum Specific Residue of Inorganic Bromide Found in Various Foods (ppm/lb/min)[a]

Maximum Specific Residue	Commodities
0-5	Baking powder, butter, chewing gum, coffee (whole roasted) macaroni, marshmallows, oleomargarine, shortening, tapioca, flour, tea, yeast (dry)
5-10	Breakfast cereals (precooked), cake mix, candy, cheese, ginger (ground), milk (dried), nutmeg (ground), pancake mix, ground red pepper, veal loaf
10-15	Cinnamon (powdered), coffee (ground roasted), cocoa
15-20	Allspice, beef cuts, gelatin, noodles, peanuts, pie crust mix
20-30	Cornmeal, cream of wheat, frankfurters, pork cuts, rice flour
30-40	Bacon, cattle feed (mixed), dry dog food, wheat flour (white and whole wheat)
40-50	Soy flour
75-100	Grated Parmesan cheese
100-125	Powdered eggs

[a] Specific Residue = $\dfrac{\text{ppm bromide increase from fumigation}}{\text{rate of fumigation (lb/min)}}$

SOURCE: Getzendaner et al. (1968). Bromide residues from methyl bromide fumigation of food commodities. Reprinted with permission from the *Journal of Agricultural and Food Chemistry* 16:265-271. Copyright by the American Chemical Society.

fumigations have indicated some potential residue levels. Bielorai and Alumot (1966) analyzed wheat and barley after treating it for 72 hours with 40 gm/m^3 of one mixture of the commercial fumigant described above. They found residues of 21.0 and 27.9 ppm in wheat and barley, respectively, on day zero, decreasing to 4.6 and 5.3 ppm, respectively, by day 17.

Alumot and Bielorai (1969) analyzed several grains fumigated with a fumigant containing chloroform and aired at different temperatures for two months. Initial chloroform residues of about 200 ppm in barley, corn, and sorghum were reduced to less than 20 ppm when ventilated at 30°C and were below detection when ventilated at 17°C. Lower fumigant dosages produced lower initial residues, which were reduced to less than 5 ppm after only 17 days of aeration.

The Panel on Fumigant Residues in Grain (1974) reported on a collaborative study of residues in wheat and maize samples taken two and seven days after fumigation with chloroform. Chloroform residues in fumigated wheat were 35±4 mg/kg and 35±5 mg/kg and in fumigated maize were 77±11 mg/kg and 74±11 mg/kg after two and seven days, respectively.

Higher residues of chloroform were found by Malone (1969) in unground materials than in ground materials. It appears that chloroform is lost during grinding and processing, but insufficient data are available to determine this for flour and other fractions. Malone suggests that chloroform may be present as a product from reaction between grain constituents and carbon tetrachloride, the latter being used more extensively than chloroform in fumigant mixtures.

ANALYSIS OF GLOBAL MASS BALANCES

The continental and marine background concentrations and the range of urban concentrations of the nonfluorinated halomethanes are shown in Table 5.16. Since only six of these compounds have ever been measured in the atmosphere, any discussion of the atmospheric budget must be restricted to these six compounds: the monohalomethanes—methyl chloride (CH_3Cl), methyl bromide (CH_3Br), and methyl iodide (CH_3I)—and methylene chloride (CH_2Cl_2), chloroform ($CHCl_3$), and carbon tetrachloride (CCl_4).

The Monohalomethanes and Methylene Chloride

The monohalomethanes are thought to originate primarily from natural sources because anthropogenic emissions of these compounds do not account for their tropospheric concentrations (Table 5.20). For several reasons, the oceans are generally believed to be the major monohalomethane source. Lovelock (1975) and Singh et al. (1976) have reported significant supersaturation of these compounds in surface ocean water and the tropospheric concentrations of these compounds generally have been found to be much higher in marine than in continental air samples.

However, it is possible that a significant portion of methyl bromide and methyl chloride results from direct anthropogenic emissions or from secondary reactions of anthropogenic precursors. The high methyl bromide values found in cities probably are due to secondary formation from the gasoline additive, ethylene dibromide (CH_2Br-CH_2Br), in combustion of gasoline since most anthropogenic bromine is used in the formation of that additive. This has been corroborated by recent measurements by Harsch and Rasmussen (1977), who found methyl bromide concentrations in excess of 50 ppb (v/v) in automobile exhausts.

Ethylene dichloride (CH_2Cl-CH_2Cl) and vinyl chloride ($CH_2=CHCl$) are possible sources for atmospheric methyl chloride. Methyl chloride has estimated global anthropogenic emissions of 2×10^{10} g/yr. Ethylene dichloride emissions are estimated at 56.5×10^{10} g/yr, and those for vinyl chloride are about 35.2×10^{10} g/yr. With any significant conversion of these compounds to methyl chloride in the atmosphere, anthropogenic sources could account for a significant portion of the estimated global methyl chloride emissions of 5.2×10^{12} g/yr that Yung et al. (1975) calculated. Preliminary panel calculations estimate a possible source strength of 1.05×10^{10} g/yr methyl chloride from cigarette smoking (see this chapter, section on Combustion Processes and Thermal Degradation). In addition, recent estimates of methyl chloride production in combustion calculate a source of 2.1×10^{11} g/yr for the United States alone (Palmer 1976), about 40 percent resulting from the burning of polyvinyl chloride (PVC) in waste and the rest from forest fires.

The reason for the anomalously high values for methyl iodide found in several urban areas is not so readily explicable, since no significant anthropogenic sources of iodinated hydrocarbons have been detected. It is possible that the high methyl iodide value found in Bayonne, New Jersey (Lillian et al. 1975) may have been due to a nearby source or to faulty analytical data. Singh et al. (1976) attribute the high values they measured in Los Angeles to "adverse meteorological conditions the first 60 hours of monitoring."

TABLE 5.20 Estimated Yearly Source Strengths of Chloroform and Three Monohalomethanes (10^{12} grams/year)

	Methyl Chloride	Methyl Bromide	Methyl Iodide	Chloroform
Total Source Strength[a]	5.2	7.7×10^{-2}	0.74 0.27	0.23 – 0.99
Global Industrial Source	7.9×10^{-3}	$0.5 – 2.0 \times 10^{-2}$	Insignificant	12.4×10^{-3}
U.S. Industrial Source (1973)	5.2×10^{-3}	1.1×10^{-2}	Insignificant	6.5×10^{-3}

[a]Estimated from current atmospheric levels.

SOURCE: From Yung et al. (1975), Wofsy et al. (1975), Liss and Slater (1974), and Panel estimates based on Table 5.1.

The large uncertainties present in natural and anthropogenic sources for the monohalomethanes make it impossible to construct an accurate atmospheric budget for them. Best Panel estimates of the source strengths required to maintain their current atmospheric levels are listed in Table 5.20. More detailed information is needed to determine the global distribution pattern and sources, both natural and anthropogenic, of these compounds.

Of the three remaining nonfluorinated halomethanes for which there are atmospheric measurements, methylene chloride has only been measured a few times in the atmosphere (Cox et al. 1976; see footnote 3) despite the fact that it is emitted by man in far greater amounts (346.4×10^9 g/yr) than any of the other compounds under consideration. The main reasons for this lack are the relatively short atmospheric life of this compound and the fact that the electron capture detector lacks sensitivity for a compound with only two chlorine atoms per molecule. The virtual lack of atmospheric measurements makes it futile to attempt any sort of mass balance for methylene chloride, despite its large anthropogenic emissions and possible importance to the total chlorine budget of the atmosphere.

The only two compounds that are produced and released by man in significant amounts and about which we have adequate information to make estimates of their atmospheric budgets--their sources, sinks, and transport mechanisms--are chloroform and carbon tetrachloride. There are reasonably good data on the concentration of chloroform in the atmosphere and the oceans, and carbon tetrachloride has been studied extensively both for its environmental distribution and its anthropogenic emissions inventory. Table 5.21 lists the best estimates available of the critical factors for determining the mass balances for the four chlorinated methanes, but the available data make it possible to attempt a detailed account only for chloroform and carbon tetrachloride.

Chloroform

As Table 5.21 shows, the background tropospheric concentration of chloroform is from 20 to 40 ppt (v/v), with higher concentrations associated with marine air and lower levels with continental air samples. Furthermore, the surface mixed layer of the oceans has nearly the same amount of chloroform as the overlying air mass. These facts point toward the marine environment as a chloroform source, although whether this is due to natural production by marine organisms or simply to the degradation of anthropogenic precursors, such as carbon tetrachloride, trichloroethylene, perchloroethylene, and 1,1,1-trichloroethane is uncertain.

TABLE 5.21 Preliminary Mass Balances for the Chlorinated Methanes

	Chloroform	Carbon Tetrachloride	Methyl Chloride	Methylene Chloride
Average Concentrations				
Troposphere (10^{-12} ml/ml air)	20-40	110-120	650-750	35
Surface ocean water (10^{-12} g/ml seawater)	10	0.4-0.5	14	–
1976 Production (10^9 g/year)				
United States	133	388	167	247
Total "free world"	260	752	272	456
1973 Annual Emissions (10^9 g/year)				
United States	6.5	35	5.2	199
Total "free world"	12.4	59	–	–
Cumulative Emissions (10^{12} g)				
United States	0.53	2.13	0.05	2.44
Total "free world"	0.63	2.52	0.08	4.44
Tropospheric Burden (10^{12} g)	0.47	2.9	5.9	0.46
Stratospheric Burden (10^{12} g)	–	0.2	–	–
Oceanic Burden (10^{12} g)	0.36	0.018	0.5	–

Chloroform also has some direct anthropogenic sources, as the elevated levels in urban areas demonstrate (Table 5.16). However, there clearly are other large sources of the compound. Direct, worldwide anthropogenic chloroform emissions are estimated at only 12.4 x 10^9 g/yr, while total global emissions are estimated by Yung et al. (1975) at between 230 to 990 x 10^9 g/yr (Table 5.20). Yung et al. suggested that the bleaching of wood pulp for paper might be a significant source of chloroform. Appleby et al. (1976a) also have claimed that the photooxidation of trichloroethylene, which has estimated emissions of 648.3 x 10^9 g/yr, might be a significant atmospheric chloroform source. Production of chloroform in gasoline combustion, indicated by the large concentration of $CHCl_3$ in car exhaust (Table 5.14), is another possible source.

A factor that may cause the apparent flux of chloroform from the ocean has been proposed by Eriksson (1959). He suggested that atmospheric gases whose solubilities in seawater are dependent on temperature would show a meridional circulation in the oceans due to lower surface seawater temperature at high latitudes and greater solubility of the gases in the ocean at these latitudes. When surface seawater from high latitudes circulates to temperate and tropical zones, and the surface mixed layer gets warmer, dissolved gases become supersaturated and diffuse from the ocean into the atmosphere.

Most measurements of high chloroform levels in marine air have been made along the coastlines in California, England, and Ireland, where there are large direct injections of pollutants into the coastal waters. It is possible that supersaturation of the surface ocean water and the elevated levels in marine air are simply the result of direct contamination by urban and industrial runoff. There is some support for this idea in the results of Murray and Riley (1973), who found higher atmospheric concentrations of chloroform at rural coastal stations in Great Britain than at stations in the northeast Atlantic. However, both they and Su and Goldberg (1976) found relatively high concentrations of chloroform (8 to 15 x 10^{-3} µg/l of sea water) in the open-ocean mixed layer of both the Atlantic and Pacific, far from any direct sources of contamination.

No one has ever made an historical emissions inventory of chloroform, as have been made recently for the fluorocarbons and carbon tetrachloride. A significant portion of the chloroform in the environment may originate in secondary formation from anthropogenic precursors and other compounds containing three or more chlorine atoms per molecule. Graedel and Allara (1976) have concluded that the formation of chloroform and other chlorinated hydrocarbons from natural atmospheric reactions is negligible. Secondary formation from anthropogenic precursors and the marine environment as chloroform sources need further investigation. As the preceding discussion indicates, the relative contribution of anthropogenic and natural sources to atmospheric concentrations of chloroform has not yet been resolved.

Carbon Tetrachloride

Information on carbon tetrachloride is more complete than for any of the other compounds under consideration and should permit the development of a detailed global budget, with consideration of sources, sinks, and modes of transport in the environment.

Four independently authored papers dealing with the history of the United States and global CCl_4 emissions have been published in the past year, and all reached the same conclusion: most, if not all, of the atmospheric carbon tetrachloride worldwide is man-made, and natural sources need not be invoked to explain its presence or distribution in the atmosphere.

Altshuller (1976) examined the production history of CCl_4 in the United States from 1907 to 1974 and estimated that cumulative emissions in the United States amount to approximately 1.7×10^{12} g (Table 5.22) whereas cumulative emissions from outside the United States were estimated at 0.1×10^{12} g, with an upper limit for total worldwide emissions ranging to 2.3×10^{12} g. Using these estimates and an infinite tropospheric lifetime for carbon tetrachloride, Altshuller arrived at an expected tropospheric concentration of 65 to 88 ppt (v/v) for carbon tetrachloride. To account for the loss of CCl_4 to the stratosphere, he assumed equal mixing ratios in the troposphere and stratosphere, which led him to predict a tropospheric concentration of 60 to 80 ppt (v/v) for carbon tetrachloride (Table 5.23). This value happens to agree reasonably well with measurements made in 1971 and 1972 by Lovelock et al. (1974) and Wilkniss et al. (1973) in the Atlantic Ocean, which ranged from 70 to 80 ppt (v/v).

Singh et al. (1976) analyzed production and usage information for the United States, Japan, and Western Europe and calaculated cumulative U.S. emissions through 1973 of CCl_4 at about 1.9×10^{12} g, while those for the rest of the world are approximately 0.6×10^{12} g (Table 5.22). Thus they arrive at cumulative worldwide CCl_4 emissions of 2.5×10^{12} g (considerably higher than Altshuller's estimate of 1.8×10^{12} g, but fairly close to his upper limit of 2.3×10^{12} g). Using these emissions levels and an atmospheric lifetime for CCl_4 of 75 years, and based upon photolysis in

TABLE 5.22 Summary of Data from Various Investigators on Carbon Tetrachloride, by Source

Carbon Tetrachloride Data		Source
Tropospheric Concentration	120 ppt v/v (1976)	Rasmussen, Stanford Research Institute (1975, 1976), Su and Goldberg (1976)
Surface Ocean Water	0.5×10^{-12} g/ml (1975)	Su and Goldberg (1976)
	0.14×10^{-12} g/ml (1973)	Murray and Riley (1973b)
	0.25×10^{-12} g/ml (1973)	Pearson and McConnell (1975)
Annual U.S. Emissions	20.8×10^9 g/yr (1973)	A.D. Little, Inc. (1975)
	40.9×10^9 g/yr (1973)	McCarthy (1974)
	41.7×10^9 g/yr (1973)	Singh et al. (1975, 1976)
	23.0×10^9 g/yr (1974)	Altshuller (1976)
	45.0×10^9 g/yr (1972)	Galbally (1976)
	27.5×10^9 g/yr (1973)	NRC (1976a)
	34.2×10^9 g/yr (1973)	Panel estimate
Annual World Emissions	41.7×10^9 g/yr (1973)	A.D. Little, Inc. (1975)
	81.8×10^9 g/yr (1972)	McCarthy (1974)
	96.4×10^9 g/yr (1973)	Singh et al. (1975, 1976)
	47.5×10^9 g/yr (1973)	Altshuller (1976)
	90.0×10^9 g/yr (1973)	Galbally (1976)
	55.0×10^9 g/yr (1973)	NRC (1976a)
	59.0×10^9 g/yr (1973)	Panel estimate, "free world"
Volume of Troposphere	3.5×10^{24} cm	Krey et al. (1976)
Tropospheric Burden	4.2×10^{14} ml (STP) = 2.9×10^{12} g (1976)	Krey et al. (1976)
Stratospheric Burden	0.18×10^{12} g (1975)	Krey et al. (1976)
Cumulative U.S. Emissions	1.9×10^{12} g (1973)	Singh et al (1975, 1976)
	1.7×10^{12} g (1974)	Altshuller (1976)
	2.1×10^{12} g (1975)	NRC (1976a)
	2.1×10^{12} g (1975)	Panel estimate
Cumulative Total World Emissions	2.5×10^{12} g (1973)	Singh et al. (1975, 1976)
	1.8×10^{12} g (1974)	Altshuller (1976)
	4.1×10^{12} g (1975)	NRC (1976a)
	2.5×10^{12} g (1975)	Panel estimate, "free world"
Cumulative CCl_4 Production from Perchloroethylene Oxidation	0.7×10^{12} g (estimated)	Singh et al. (1975, 1976)
Atmospheric Lifetime	30- 50 years	Molina and Rowland (1974)
	60-100 years	Singh et al. (1975, 1976)
	18- 24 years	Krey et al. (1976)
	25- 70 (37) years	Galbally (1976)
	60 years	NRC (1976a)

NOTE: Complete references to footnotes are listed in Chapter 5, Notes.

TABLE 5.23 Tropospheric Carbon Tetrachloride Concentrations (ppt)

		References
Calculated Tropospheric Concentrations (10^{-12} ml/ml)	72 – 95 90 ± 40 60 – 88 120 – 162	Singh et al. (1976) Galbally (1976) Altshuller (1976) NRC (1976a)
Measured Tropospheric Concentrations (10^{-12} ml/ml)	130 ± 10 120 ± 10 125 ± 10 130 – 170 65 ± 4 150 ± 20 60 59 – 65	Rasmussen (1976)[4] Su and Goldberg (1976) Singh et al. (1977) Lovelock (1976)[5] Penkett (1976)[6] Pearman (1976)[7] Goldan (1976)[8] Sandalls (1976)[9]

NOTE: Complete references to footnotes are listed in Chapter 5, Notes.

the stratosphere as the only significant sink, Singh et al. calculate an estimated tropospheric concentration of 72 to 95 ppt (v/v), which agrees with Lovelock's 1971-1972 measurement of 71 ppt (v/v).

In addition, Singh et al. estimate that the photooxidation of perchloroethylene ($CCl_2=CCl_2$) could be an important secondary source of CCl_4. On the basis of their preliminary work, they estimate that as much as 8 percent of all $CCl_2=CCl_2$ emitted may be converted to carbon tetrachloride, enough to produce 0.7×10^{12} g CCl_4. They arrive at a possible total of 3.2×10^{12} g of CCl_4 attributable to man-made sources, compared to 3.1×10^{12} g, the best current estimate of the total atmospheric carbon tetrachloride burden (Table 5.22).

Galbally (1976) also analyzed worldwide carbon tetrachloride production and emissions data from 1914 to 1973 and estimated an expected global average concentration of 90±40 ppt (v/v) on the basis of four sinks: gas phase reactions, hydrolysis in the oceans, uv photolysis in the stratosphere, and removal in biological systems. His predicted atmospheric concentration levels are compared with the expected levels due to man-made emissions of CCl_4 in Figure 5.6. Measured values generally fall within the uncertainty limits of his predictions. He also calculated the expected steady state concentration of carbon tetrachloride, based on the assumption that all future CCl_4 emissions will remain at or below present levels. He predicts a final equilibrium global average concentration for CCl_4 of 280 ppt (v/v) or less.

Finally, the National Academy of Sciences report (NRC 1976a) on the effects of halocarbons on stratospheric ozone, which assumed that total world carbon tetrachloride production and emissions are exactly double that of the United States, arrived at a figure for cumulative worldwide emissions of carbon tetrachloride considerably higher than other investigators: 4.1×10^{12} g (Table 5.22). Using an atmospheric lifetime of 60 years for CCl_4, the report estimated that 75 percent of this amount is still in the atmosphere. This would give an average tropospheric concentration of 120 ppt (v/v), which is higher than the estimates of Altshuller, Singh et al., and Galbally but compares favorably with recent measurements by Singh et al. (1977) of 115 to 135 ppt (v/v) and by Rasmussen[*] of 120 to 140 ppt (v/v).

It is clear, in looking at all the relevant data on carbon tetrachloride in Table 5.22 that significant uncertainties exist in a number of areas. Despite the fact that their total cumulative global emissions differ by almost 50 percent, Singh et al. and Altshuller managed to

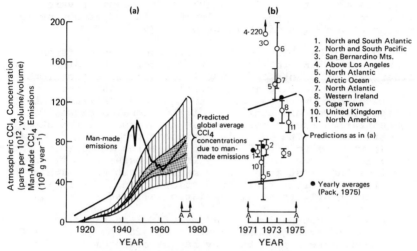

(a) The inner hatched area indicates the uncertainty due to loss mechanisms. The shaded area indicates the uncertainty due to loss mechanisms and a = 30 percent variation in the emission rates.

(b) Observed CCl_4 concentrations and predicted global average concentrations from (a).

SOURCE: Galbally (1976) Man-made carbon tetrachloride in the atmosphere. *Science* 193:573-576. Copyright by the American Association for the Advancement of Science.

FIGURE 5.6 Galbally's predicted global average CCl_4 concentration due to man-made emissions (a) and observed CCl_4 concentrations (b).

arrive at apparently fortuitous agreement with each other and with the 1971-1972 measurements by Lovelock (1974) by the way they handled the question of CCl_4 atmospheric sinks (Table 5.23). Altshuller assumed an atmospheric CCl_4 lifetime of several hundred years and ignored the question of atmospheric sinks, while Singh et al. (1975) included only the photolysis of CCl_4 in the stratosphere as a sink of any significance. By assuming an average CCl_4 atmospheric lifetime of 75 years, Singh et al. arrived at an estimated average tropospheric concentration of 72 to 74 ppt (v/v), which is close to Lovelock's 1971-1972 mid-ocean value of 71.2 ppt (v/v) and to Altshuller's estimated tropospheric CCl_4 concentration of 60 to 80 ppt (v/v).

On the other hand, Galbally includes four different sinks, estimating their relative importance by using first order rate constants of 1×10^{-3} for gas phase reactions, 7×10^{-3} for hydrolysis in the oceans, 1.7×10^{-2} for stratospheric UV photolysis, and 2×10^{-3} for removal by the biosphere. But his estimates of the global average concentration are no more precise than those of the other investigators. The Galbally estimates range from 50 to 130 ppt (v/v), due to a ± 30 percent variation in emission rates and a ± 50 percent uncertainty in the loss rate constants. The fact that most atmospheric measurements fall within the uncertainty limits of his predictions is not surprising, given the nearly threefold variation allowed.

Finally, the National Academy of Sciences' estimate of a tropospheric average concentration of 120 ppt (v/v) for CCl_4 is in good agreement with many recent measurements, but their assumption that total cumulative global emissions are twice those of the United States is at odds with the calculations of Altshuller, Singh et al. (1976) and Galbally, all of whom estimate that the United States is responsible for at least 75 percent of total global emissions.

Two problems that prevent resolution of this conflict are evident in Table 5.22: the uncertainties that still exist about the tropospheric concentration and the atmospheric lifetime of CCl_4. Although most recent tropospheric measurements indicate concentrations of CCl_4 in the range 115 to 140 ppt (v/v), there are also several groups currently reporting carbon tetrachloride levels roughly half as large, 60 to 65 ppt (v/v) (Table 5.23). Until this disagreement can be resolved, it will be difficult to determine exactly what aspects of the models need revision. Presently, one could make equally valid assertions that anthropogenic sources are or are not sufficient to explain the atmospheric distribution of CCl_4, or that the sinks used in various models are too large or too small to account for current levels of atmospheric carbon tetrachloride. Obviously, a program of intercalibration among the various laboratories making CCl_4 measurements is needed to eliminate the discrepancies; some absolute calibration must be established as soon as possible.

Once the actual, as opposed to estimated, concentration of carbon tetrachloride in the troposphere has been determined, then the question of the atmospheric lifetime of CCl_4 must be addressed. As can be seen in Table 5.25, the estimates of the atmospheric lifetime of CCl_4 range from 18 to 100 years. The estimates based on photolysis by uv radiation in the stratosphere as the only significant sink agree fairly well on 60 to 75 years. Considering that CCl_4 has been manufactured only for the past 70 years, there is little difference in the concentrations predicted by this range of atmospheric lifetimes. However, the much shorter atmospheric lifetime estimated by Krey et al. (1976) on the basis of stratospheric CCl_4 measurements and the lifetime of about 37 years estimated by Galbally on the basis of four sinks (with hydrolysis in the oceans being the major one after uv photolysis) both predict much lower levels of CCl_4 in the atmosphere if the same emissions are assumed.

This fact is underscored by Pierotti and Rasmussen (1976), whose results indicate that CCl_4 flux to the oceans may have a rate constant of 1.7×10^{-2}, equal to the stratospheric photolysis rate and more than twice the rate assumed by Galbally for hydrolysis in the ocean. The exact mechanism involved in this oceanic sink is uncertain, but it may stem from the bioaccumulation of carbon tetrachloride in marine organisms, and then removal of CCl_4 by the deposition of organic particulates to deeper water, in much the same way that compounds such as the PCBs are believed to be removed from the surface water layer.

There are a number of research areas which should be pursued to settle the uncertainties now existing about carbon tetrachloride. In addition to the absolute calibration needed to establish the accurate CCl_4 tropospheric concentration and better estimates of the atmospheric lifetime, there are specific experiments that would help to settle the questions of the CCl_4 sources and sinks.

First, the carbon-14 specific activity of CCl_4 should be measured to determine if CCl_4 is entirely anthropogenic or if it has some natural sources. If the activity were very low, this would indicate that CCl_4 is a product of industrial synthesis from petroleum hydrocarbons. If it were near the atmospheric value for carbon-14 specific activity, this would indicate a natural formation mechanism.

The same approach could be applied to any of the halogenated hydrocarbons whose origin is in question.

Second, a number of experiments could be carried out to determine if the oceans are a significant sink for carbon tetrachloride. The CCl_4 concentration in phytoplankton and zooplankton should be measured to determine if there is significant biomagnification. The concentration in the deep ocean below the surface mixed layer and in marine sediments should be measured to determine if these areas are a sink, as evidence suggests.

Finally, the concentration of carbon tetrachloride and the other halogenated hydrocarbons should be measured in glaciers to determine their atmospheric history for the past several hundred years. This history would indicate if the current atmospheric levels have been present for a long time or if they have shown significant increases during the twentieth century, which would indicate their anthropogenic contributions.

With these experiments and with the systematic measurement of the nonfluorinated halomethanes in the atmosphere, we would be able to construct a more definitive picture of the global atmospheric cycle of these compounds. Our knowledge of these compounds in the environment has grown enormously in the past 10 years, and it is important to continue research in this area even though no immediate environmental problem is evident. After all, when Lovelock first began measuring the fluorocarbons, who would have thought they would become a matter of urgent concern as a possible threat to the environment?

NOTES

1. Personal communication from J. Harrald (1976) U.S. Coast Guard.

2. Personal communication from J. Morris (1976) E.I. du Pont de Nemours and Company, Wilmington, Delaware.

3. Backus, J. (undated) Flammability of Urethane Foams. Pittsburgh, Pa.: Mobay Chemical Co.

4. Personal communication from R.A. Rasmussen (1976) Washington State University.

5. Personal communication from J.E. Lovelock (1976) Reading University, Reading, England.

6. Personal communication from S. Penkett (1976) AERE Harwell, Oxfordshire, England.

7. Personal communication from G. Pearman (1976) CSIRO, Victoria, Australia.

8. Personal communication from P. Goldan (1976) NOAA, Boulder, Colorado.

9. Personal communication from F.J. Sandalls (1976) AERE Harwell, Oxfordshire, England.

REFERENCES

Altshuller, A.P. (1976) Average tropospheric concentration of carbon tetrachloride based on industrial production, usage, and emissions. Environmental Science and Technology 10(6):596-598.

Alumot, E. and R. Bielorai (1969) Residues of fumigant mixtures in cereals fumigated and aired at two different temperatures. Journal of Agricultural and Food Chemistry 17:869.

Appleby, A., J. Kazazis, D. Lillian, and H.B. Singh (1976a) Atmospheric formation of chloroform from trichloroethylene. Journal of Environmental Science and Health 12:711-715.

Appleby, A., D. Lillian, and H.B. Singh (1976b) Comment on "Atmospheric Halocarbons." Geophysical Research Letters 3(4):237.

Arthur D. Little, Inc. (1975) Preliminary Economic Impact Assessment of Possible Regulatory Action to Control Atmospheric Emissions of Selected Halocarbons. EPA-450/3-75-073; NTIS No. PB-247 115. Springfield, Va.: National Technical Information Service.

Bellar, T.A., J.J. Lichtenberg, and R.C. Kroner (1974) The Occurrence of Organohalides in Chlorinated Drinking Waters. EPA-670/4-74-008. Washington, D.C.: U.S. Environmental Protection Agency.

Berck, B. (1974) Fumigant residues of carbon tetrachloride, ethylene dichloride and ethylene dibromide in wheat, flour, bran, middlings, and bread. Journal of Agricultural and Food Chemistry 22:977-984.

Berck, B. (1975) Analysis of fumigants and fumigant residues. Journal of Chromatographic Science 13:256-266.

Bielorai, B. and E. Alumot (1966) Determination of residues of a fumigant mixture in cereal grain by electron-capture gas chromatography. Journal of Agricultural and Food Chemistry 14:622.

Bielorai, B. and E. Alumot (1975) The temperature effect on fumigant desorption from cereal grain. Journal of Agricultural and Food Chemistry 23:426-429.

Bondi, A. and E. Alumot (1972) As reported by E.E. Kenaga, 1971, IUPAC meeting. WHO Pesticide Residues Series, No. 1, 1972. 1971 Evaluations of Some Pesticide Residues in Food. Geneva: The Monographs, World Health Organization.

Burreson, B.J., R.E. Moore, and P.P. Roller (1975) Haloforms in the essential oil of the alga Asparagopsis taxiformis (Rhodophyta). Tetrahedron Letters 7:473-476.

Chopra, N.M. and L.R. Sherman (1972) Systematic Studies on the Breakdown of p,p'-DDT in Tobacco Smokes. Analytical Chemistry 44:1036-1038.

Coleman, W.E., R.D. Lingg, R.G. Melton, and F.C. Kopfler (1976) The occurrence of volatile organics in five drinking water supplies using gas chromatography/mass spectrometer. Page 305, Identification and Analysis of Organic Pollutants in Water, edited by L. Keith. Ann Arbor, Mich.: Ann Arbor Science Publishers, Inc.

Cox, R.A., R.G. Derwent, A.E.J. Eggleton, and J.E. Lovelock (1976) Photochemical oxidation of halocarbons in the troposhere. Atmospheric Environment 10(4):305-308.

Cronn, D.R., R.A. Rasmussen, and E. Robinson (1976) Measurement of troposheric halocarbon by gas chromatography-mass spectrometry. Report for Phase I of EPA Grant No. R0804033-01. National Environmental Research Center, Research Triangle Park, N.C. August 23.

Dennis, N.M., G. Eason, and H.B. Gillenwater (1972) Formation of methyl chloride during methyl bromide fumigations. Journal of Economic Entomology 65:1753-1754.

Derby, J.V. and R.W. Freedman (1974) Vapor-phase pyrolysis and gas chromatography analysis of fire retardant materials containing polyvinyl chloride. American Laboratory 6(5):10-12, 14, 16.

Eriksson, E. (1959) Pages 147-157, The Atmosphere and the Sea in Motion, edited by B. Bolin. New York: The Rockefeller Institute Press.

Galbally, I.E. (1976) Man-made carbon tetrachloride in the atmosphere. Science 193:573-6.

Getzendaner, M.E. and H.H. Richardson (1966) Bromide residues from methyl bromide fumigation of fruits and vegetables subjected to quarantine schedules. Journal of Agricultural and Food Chemistry 14:59-62.

Getzendaner, M.E., A.E. Doty, E.L. McLaughlin, and D.L. Lindgren (1968) Bromide residues from methyl bromide fumigation of food commodities. Journal of Agricultural and Food Chemistry 16:265-271.

Graedel, T.E. and D.L. Allara (1976) Tropospheric halocarbons: estiamtes of atmospheric chemical production. Atmospheric Environment 10(5):385-388.

Grimsrud, E.P. and R.A. Rasmussen (1975) Survey and analysis of halocarbons in the atmosphere by gas chromatography-mass spectrometry. Atmospheric Environment 9:1014-1017.

Hampel, C.A. and G.G. Hawley, eds. (1973) The Encyclopedia of Chemistry, 3rd edition. New York: Van Nostrand Reinhold.

Harsch, D. (1977) Study of halocarbon concentrations in indoor environments. Final report for EPA Contract No. WA 6-99-2922-J. Washington, D.C.: U.S. Environmental Protection Agency.

Harsch, D. and R.A. Rasmussen (1977) Identification of methyl bromide in urban air. EPA Contract No. WA-6-99-2922-J. Office of Research and Development. Washington, D.C.: U.S. Environmental Protection Agency. (Unpublished report)

Johnson, W.R., R.W. Hale, J.W. Nedlock, H.J. Grubbs, and D.H. Powell (1973) The distribution of products between mainstream and sidestream smoke. Tobacco Science 17:141-144.

Kenaga, E.E. (1957) Evaluation of grain fumigants. Down to Earth (Winter) 1957.

Krey, P.W., R. Lagomarsino, M. Schonberg, and I.E. Toonkel (1976) CCl_4 in Ground Level Air and Stratosphere. Health and Safety Laboratory Environmental Quarterly, April 1, 1976, I-21-I-49.

Lillian, D., H.B. Singh, A. Appleby, L. Lobban, R. Arnts, R. Gumpert, R. Hague, J. Toomey, J. Kazazis, M. Antell, D. Hansen, and B. Scott (1975) Atmospheric fates of halogenated compounds. Environmental Science and Technology 9(12):1042-1048.

Liss, P.S. and P.G. Slater (1974) Flux of gases across the air-sea interface. Nature 247:181-184.

Lovelock, J.E. (1974) Atmospheric halocarbons and stratospheric ozone. Nature 252:292-294.

Lovelock, J.E. (1975) Natural halocarbons in the air and in the sea. Nature 256:193-194.

Lovelock, J.E., R.J. Maggs, and R.J. Wade (1973) Halogenated hydrocarbons in and over the Atlantic. Nature 241:194-196.

Lovelock, J.E., R.J. Maggs, and R.J. Wade (1974) Halogenated hydrocarbons in and over the Atlantic. Nature 247:194-196.

Lynn, G.E. and F.A. Vorches, Jr., eds. (1957) Residues in food and feeds resulting from fumigation of grains with the common liquid formulations of carbon disulfide, carbon tetrachloride, ethylene dichloride, and ethylene dibromide. Journal of Association of Official Agricultural Chemists 40:163-209.

Malone, B. (1969) Analysis of grains for multiple residues of organic fumigants. Journal of Association of Official Agricultural Chemists 52:800-805.

McCarthy, R.L. (1974) Fluorocarbons in the environment. Paper presented before the National Geophysical Union, San Francisco, California, December 31, 1974. Wilmington, Del.: E.I. duPont de Nemours and Company.

McConnell, G., D.M. Ferguson, and C.R. Pearson (1975) Chlorinated hydrocarbons and the environment. Endeavour 34(121):13-18.

McKinney, J.D., R.R. Maurer, J.R. Hass, and R.O. Thomas (1976) Possible factors in the drinking water of laboratory animals causing reproductive failure. Page 417, Identification and Analysis of Organic Pollutants in Water, edited by L. Keith. Ann Arbor, Mich.: Ann Arbor Science Publishers, Inc.

McMahon B. (1971) Analysis of commercially fumigated grains for residues of organic fumigants. Journal of the Association of Official Analytical Chemistry 54:964-965.

Molina, M.J. and F.S. Rowland (1974) Predicted present stratospheric abundances of chlorine species from photodissociation of carbon tetrachloride. Geophysical Research Letters 1(7):309-312.

Moore, R.E., F.X. Woolard, and P.P. Roller (1976) Halogenated acetamides, acetic acids, acrylic acids, but-3-en-2-ols, and isopropanols from the red alga

Asparagopsis taxiformis. Paper presented at 1976 meeting of the American Chemical Society, San Francisco, California.

Morris, J.C. and G. McKay (1975) Formation of Halogenated Organics by Chlorination of Water Supplies. EPA-600/1-75-002; NTIS No. PB 241-511. Springfield, Va.: National Technical Information Service.

Murray, A.J. and J.P. Riley (1973) Occurrence of some chlorinated aliphatic hydrocarbons in the environment. Nature 242:37-38.

National Research Council (1976a) Halocarbons: Effects on Stratospheric Ozone. Washington, D.C.: National Academy of Sciences.

National Research Council (1976b) Halocarbons: Environmental Effects of Chlorofluoromethane Release. Appendix D, Sources of CFM Emissions and Approaches to Their Reduction. Washington, D.C.: National Academy of Sciences.

Neely, W.B. (1976) Material balance analysis of trichlorofluoromethane and carbon tetrachloride in the atmosphere. Health and Environmental Research. Midland, Mich: The Dow Chemical Co. (Unpublished manuscript)

Neely, W.B., G.E. Blau, and T. Alfrey, Jr. (1976) Mathematical models predict concentration-time profiles resulting from chemical spill in a river. Environmental Science and Technology 10:72-76.

Newsome, J.R., V. Norman, and C.H. Keith (1965) Vapor phase analysis of tobacco smoke. Tobacco Science 9:102-110.

Ohta, T., M. Morita, and I. Mizoguchi (1976) Local distribution of chlorinated hydrocarbons in the ambient air in Tokyo. Atmospheric Environment 10(7):557-560.

Paciorek, K.L., R.H. Kratzer, J. Kaufman, and A.M. Hartstein (1974a) Oxidative thermal degradation of selected polymeric compositions. Journal of American Industrial Hygiene Association 35:175-180.

Paciorek, K.L., R.H. Kratzer, J. Kaufman, and J. Nakahara (1974b) Oxidative thermal decomposition of poly(vinyl chloride) compositions. Journal of Applied Polymer Science 18:3723-3729.

Palmer, T.Y. (1976) Combustion sources of atmospheric chlorine. Nature 263:44-46.

Panel on Fumigant Residues in Grain (1974) The determination of residues of volatile fumigants in grain. Analyst 99:570-576.

Pearson, C.R. and G. McConnell (1975) Chlorinated C_1 and C_2 hydrocarbons in the marine environment. Proceedings of the Royal Society of London, Series B 189(1096):305-332.

Philippe, R.J. and M.E. Hobbs (1956) Some components of the gas-phase of cigarette smoke. Analytical Chemistry 28:2002-2006.

Pierotti, D., R.A. Rasmussen, J. Krasnec, and B. Halter (1976) Trip Report on the Cruise of the R/V Alpha Helix from San Diego, California to San Martin, Peru. NSF Grant No. OCE 75-04688 A03. Washington, D.C.: National Science Foundation.

Pierotti, D. and R.A. Rasmussen (1976) Interim Report on the Atmospheric Measurement of Nitrous Oxide and the Halocarbons. NASA Grant NSG 7214. Washington, D.C.: National Aeronautical and Space Administration.

Roehm, R.S., S.A. Shrader, and V.A. Stenger (1943) Journal of Dairy Science 26:205.

Scudamore, K.A. and S.G. Heuser (1970) Residual free methyl bromide in fumigated commodities. Pesticide Science 1:14-17.

Scudamore, K.A. and S.G. Heuser (1973) Determination of carbon tetrachloride in fumigated cereal grains. Pesticide Science 4(1):1-12.

Seo, S.T., J.W. Balook, A.K. Burditt, Jr., and K. Ohinata (1970) Residues of ethylene dibromide, methyl bromide, and ethylene chlorobromide resulting from fumigation of fruits and vegetables infested with fruit flies. Journal of Economic Entomology 63:1093-1097.

Singh, H.B., D. Lillian, A. Appleby, and L. Lobban (1975) Atmospheric formation of carbon tetrachloride from tetrachloroethylene. Environmental Letters 10(3):253-256.

Singh, H.B., D.P. Fowler, and T.O. Peyton (1976) Atmospheric carbon tetrachloride: another man-made pollutant. Science 192:1231-1234.

Singh, H.B., L. Salas, A. Crawford, R.J. Hanst, and J.W. Spence (1977) Urban- non-urban relationships of halocarbons, SF_6, N_2O and other atmospheric constituents. Atmospheric Environment 11(9):819.

Smith, J.R. and J.C. Suggs (1976) Smoke Composition. Proceedings of the International Symposium on Air Quality and Smoke from Urban and Forest Fires, October 24-26, 1973, Fort Collins, Colorado. Washington, D.C.: National Academy of Sciences.

Stanford Research Institute (1975, 1976) Fluorinated hydrocarbons - salient statistics. Chemical Economics Handbook, Series 658, 2030. September 1975 and Supplemental Data, December 1976. Menlo Park, Calif.: Stanford Research Institute. (This is a client-private service of SRI's Chemical Information Services Department; by subscription only.)

Su, C. and E.D. Goldberg (1976) Environmental concentrations and fluxes of some halocarbons. Pages 353-374, Marine Pollutant Transfer, edited by H.L. Windom and R.A. Duce. Lexington, Mass.: Lexington Books, D.C. Heath and Company.

Symons, J.M., T.A. Bellar, J.K. Carswell, J. DeMarco, K.L. Kropp, G.G. Robeck, D.R. Seeger, C.J. Slocum, B.L. Smith, and A.A. Stevens (1975) National organics reconnaissance survey for halogenated organics (NORS). Journal of the American Water Works Association 67:634-647.

Thrune, R.I. (1963) Gases released when fire resistant epoxy resins are burned. Journal of American Industrial Hygiene Association 24:475.

U.S. Environmental Protection Agency (1975) Preliminary Assessment of Suspected Carcinogens in Drinking Water, and Appendices. A Report to Congress. Washington, D.C.: U.S. Environmental Protection Agency.

U.S. International Trade Commission (1976) Preliminary Report on U.S. Production of Selected Synthetic Organic Chemicals, May, June, and Cumulative Totals, 1976. Synthetic Organic Chemicals Series C/P-76-5, August 18, 1976. Washington, D.C.: International Trade Commission.

Wilkniss, P.E., R.A. Lamontagne, R.E. Larson, J.W. Swinnerton, C.R. Dickson, and T. Thompson (1973) Atmospheric trace gases in the Southern Hemisphere. Nature (Physical Science) 245:45-47.

Wit, S.L. (1972) Residues of insecticides in cereals and related products incorporated in the Netherlands, 1964/1966. 1971 Evaluations of Some Pesticide Residues in Food, Monograph, WHO Pesticide Residues Series, No. 1. Geneva: The Monographs, World Health Organization.

Wit, S.L., A.F.H. Besemer, H.A. Das, W. Goedkoop, F.E. Loosjes, and E.R. Meppelink (1972) Results of an investigation on the regression of three fumigants (carbon tetrachloride, ethylene dibromide, and ethylene dichloride) in wheat during processing to bread. 1971 Evaluations of Some Pesticide Residues in Food, Monograph, WHO Pesticide Residues Series, No. 1. Geneva: The Monographs, World Health Organization.

Wofsy, S.C., M.B. McElroy, and Y.L. Yung (1975) The chemistry of atmospheric bromine. Geophysical Research Letters 2:215.

Yung, Y.L., M.B. McElroy, and S.C. Wofsy (1975) Atmospheric halocarbons: a discussion with emphasis on chloroform. Geophysical Research Letters 2(9):397-399.

CHAPTER 6

EFFECTS

DIRECT ECOSYSTEM EXPOSURE AND EFFECTS

Data on the direct ecological effects of nonfluorinated halomethanes are scarce. Typical of the results yielded by extensive search in EPA files, study of chemical abstracts, and discussions with government and academic scientists is the following statement from the wildlife toxicologist in EPA's Pesticides Program:

"To date, (April 1976) there are no published reports on the toxicity of either methyl bromide or carbon tetrachloride for any wildlife species and apparently none for fish. Field hazard evaluations for fish and wildlife are also lacking. Since methyl bromide is used as a fumigant in enclosed areas, it is reasonable that few, if any, fish or wildlife poisoning incidents have ever occurred. The high vapor pressure of carbon tetrachloride and methyl bromide precludes highly toxic short term dosages from occurring on grain fed upon by wild birds. Any potential for mutagenicity or carcinogenicity in wildlife from low levels in such grain has not been studied" (Richard K. Tucker, Ecological Effects Branch, Criteria and Evaluation Division, U.S. EPA, personal communication, April 1976).

Some reasons for this scarcity of data can be inferred from the discussion in Chapter 3 of physical and chemical properties of the nonfluorinated halomethanes. The compounds are quite volatile and do not accumulate in terrestrial or aquatic environments; they are diluted rapidly to low or very low concentrations in the troposphere, where all but the completely halogenated compounds are degraded as a result of photochemical reactions. Even where major spills have occurred, ecosystem effects either have not been observed or have not been considered significant enough to warrant study, and apparent acute effects have been minimal. Consequently, significant chronic effects on fish and wildlife from long-term exposures at low levels are unlikely.

Such considerations cannot, however, entirely allay concern since laboratory and field experiments indicate that, under certain conditions, the compounds have toxic potential. Results from mammalian laboratory studies, valuable for predicting effects on mammalian wildlife, are discussed briefly in this chapter in the section on General Toxicity, and in more detail in Appendix B. Other studies concerned with toxic potential are those of McKee and Wolf (1971), and Jones (1947a, 1947b) on the toxicity of chloroform to fish. These investigators found that sticklebacks avoid solutions of 100 to 200 mg/l of chloroform in tap water, and become anesthetized at 500 mg/l. The toxicity of methylene chloride to fish has been studied, using both continuous flow and static testing techniques, by H.C. Alexander, W.M. McCarty, and E.A. Bartlett (Dow Chemical Company, Midland, Mich. 1977, unpubished data), who reported 96 hour LC_{50} values of 193 mg/l and 310 mg/l respectively. Most fish exposed recovered upon transfer to fresh water. It is important to assure that conditions that might lead to realization of these toxic potentials do not occur.

Methyl bromide, carbon tetrachloride, and to a lesser extent, chloroform are used as pesticides--basically grain and soil fumigants. Obviously, these compounds have direct effects on the organisms against which they are used; however, unintended ecosystem effects have not been documented, although non-target soil organisms are undoubtedly affected. Chapter 5 discusses the concentrations of fumigant residues in food, but levels usually are too low to cause plant or wildlife problems.

The interaction of methylene chloride, chloroform, and carbon tetrachloride with anaerobic organisms has been studied by several authors (Bauchop 1967, Jackson and Brown 1970, Sanwick and Foulkes 1971, Stickley 1970, Thiel 1969). They found inhibition in various digestive systems--sewage, sludge, rumen of cow--occurring at concentrations from approximately 1 to 100 mg/l. Bauchop, Sanwick and Foulkes, and Thiel found a 50-percent inhibition of anaerobic digestion at 1 mg/l of chloroform.

Wood et al. (1968) have shown that low concentrations of methylene chloride, chloroform, or carbon tetrachloride inhibit cobamide-dependent methyl transfer reactions in extracts of <u>Methanobacillus omelianskii</u> and the N^5-methyl tetrahydrofolate-homocysteine transmethylase of <u>Escherichia coli</u> B. According to J.M. Wood (Freshwater Biological Institute, Navarre, Minnesota, personal communication, November 1976), the possibility that the increasing levels

of low molecular weight halocarbons in natural waters and
airs might individually or additively inhibit fermentation
processes (natural, in sewage digestion plants, or in the
production of alcoholic beverages) should be investigated,
though there is no imminent hazard. Although the levels of
these materials in the environment are still markedly low--
nanograms per cubic meter in air and nanograms per liter in
water--the halocarbons in combination may still affect the
methyl transfer processes in small but detectable ways.

EFFECTS ON THE STRATOSPHERE
AND SUBSEQUENT BIOLOGICAL IMPACT

This section deals with indirect effects of halomethanes
in the atmosphere, effects that result from interactions of
these compounds with stratospheric ozone and infrared
radiation from the earth's surface. The impact of the
resulting indirect effects on both humans and ecosystems are
discussed together.

These phenomena have been considered and evaluated in
two National Academy of Sciences reports released in
September, 1976: Halocarbons: Environmental Effects of
Chlorofluoromethane Release (NRC 1976a) (the Tukey report),
and Halocarbons: Effects on Stratospheric Ozone (NRC 1976b)
(the Gutowsky report). Although the major emphases of these
reports were on the releases of chlorofluoromethanes (CFMs)
to the atmosphere and the possible resulting effects on
humans, plants, and animals, the nonfluorinated halomethanes
also were considered in some detail and their effects
compared to those of the CFMs. Because of the full and
comprehensive analyses in the Tukey and Gutowsky reports,
the Panel believed it was unnecessary to undertake
additional analyses, and the bulk of the following
discussion on effects on the stratosphere and subsequent
biological impacts is based on material in those reports,
primarily the Tukey report. Interpretations of that
material, however, are the responsibility of the Panel.

The Influence of Halomethanes on Ozone

After relatively rapid tropospheric mixing of
halogenated methanes, there is slower "leakage" into the
stratosphere and more or less random ascent to where there
is ultraviolet light in the range of 185 to 225 nanometers
(nm). In this range, ozone formation occurs, usually at
heights above 25 km. The entire process requires several
decades for the average molecule.

Chlorine-containing halomethanes absorb ultraviolet (uv)
light in the range of 185 to 225 nm and form "odd chlorine,"

initially one odd chlorine for one halomethane, but
eventually one for each chlorine in the halomethane
molecule. A small amount of the chlorine will leave the
stratosphere almost immediately, but most will remain for a
few years. Once it reaches the troposphere it will be
"rained out," mainly as HCl.

Chlorine atoms (one of a variety of chlorine-containing
species collectively called "odd chlorine") in the
stratosphere cause the destruction of ozone. In the course
of its residence, each odd chlorine destroys an average of
thousands of ozone molecules; the number depends on the
rates of various chemical actions and the rate of return to
the troposphere.

Anthropogenic and natural halogenated hydrocarbons--
including fluorocarbons 11 and 12, carbon tetrachloride,
and, to a lesser extent, fluorocarbon 22 used largely for
refrigeration, methyl bromide employed in agricultural
fumigation, and a number of hydrogen-containing and
partially halogenated chlorocarbons--may contribute to the
reduction in stratospheric ozone. It is estimated that for
the partially halogenated chloromethanes, at current release
rates, the ultimate or "steady-state" ozone reduction that
will be caused by each is relatively minor, less than 0.05
percent. The present total reduction in stratospheric ozone
by HCl, CH_3Cl, and CCl_4 (from whatever sources) is
calculated to be less than 1 percent. These three compounds
now contribute roughly the same amount of chlorine to the
stratosphere as do the CFMs, and the ozone reduction they
produce is also comparable. The importance of HCl is
reduced, however, because of its rapid "washout" from the
troposphere by rain and that of CH_3Cl by destruction
processes in the troposphere. Because of large past
emission rates of CCl_4, its estimated current ozone
reduction rate, in contrast, is appreciable--about 0.5
percent. However, since the total CCl_4 atmospheric burden
is now about 40 times the known annual injection rate, it is
plausible to assume it is near the steady state and will not
grow markedly unless emissions increase.

It has been suggested that the injection into the
atmosphere of large amounts of chlorine compounds from
natural sources, such as HCl, CH_3Cl, and perhaps CCl_4,
reduces the significance of man-made sources, but that has
not been borne out by close scrutiny. Most important is the
fact that reduction in stratospheric ozone by chlorine
compounds from man-made sources is increasing and is in
addition to whatever is caused by chlorine compounds from
natural sources. While the latter are already at their
steady-state amounts, continued release of CFMs at recent
rates will probably cause atmospheric concentrations of
chlorine, and ozone reduction, to increase tenfold or more.

All the evidence examined indicates that the long-term release of fluorocarbon 11 and fluorocarbon 12 at present rates will cause an appreciable reduction in the amount of stratospheric ozone. Specifically, it appears that their continued release at 1973 production rates would cause the ozone to decrease steadily until a probable reduction of about 6.0 to 7.5 percent is reached; these figures have an uncertainty of at least 2 to 20 percent, using what are believed to be about 95 percent confidence limits. The time required for the reduction to attain half of this steady-state value (3.0 to 3.75 percent) would be about 40 to 50 years.

The estimated magnitude of this effect from the nonfluorinated anthropogenic halomethanes, at current release rates, is not as great. But if the rate of emissions were to increase, the nonfluorinated halomethanes could become a concern.

Climatic Effects

There may be direct or indirect climatic effects from even very small concentrations of CFMs and other halogenated methanes in the atmosphere. Direct effects were evaluated by Ramanathan (1975), who considered the impact of the overall thermal energy balance in the earth's atmosphere due to the buildup in concentration of both CFMs and chlorocarbons, including CCl_4, $CHCl_3$, CH_2Cl_2, and CH_3Cl. These compounds have strong infrared (IR) bands that can absorb radiation from the earth's surface and emit it to warm the lower layers of the earth's atmosphere. This "greenhouse" effect would tend to increase the surface and atmospheric temperature. According to Ramanathan, if the concentrations of fluorocarbon 11 and fluorocarbon 12 were each increased to 2 ppb, the mean global surface temperature, under the assumption of his simplified model, could increase by as much as 0.9°C.

Based on the currently measured atmospheric concentrations of chlorinated methanes, their combined "greenhouse" effects are probably substantially less than from fluorocarbons 11 and 12. For example, Wang et al. (1976) have calculated that for atmospheric concentrations near the earth's surface of 10×10^{-4} and 2×10^{-4} ppm (v/v) for CH_3Cl and CCl_4, respectively, the combined "greenhouse" effect would be of the order of 0.01 to 0.02°C.

Indirect climatic effects arise from the possible destruction and redistribution of ozone by the odd-chlorine products that result from the photolytic decomposition of halomethanes. When ozone is removed, more uv and visible radiation reaches the ground, which tends to warm the lower

atmosphere and the earth's surface. At the same time, the
loss of uv absorption by ozone in the stratosphere reduces
heating there and less IR radiation is emitted towards the
ground. This tends to cool the lower atmosphere and the
earth's surface. Hence, the two effects compete.

Using models, the Tukey report calculated the likely
effect of CFMs on global average temperatures. The report
concluded that the estimated effects of CFM releases at 1973
rates would be sufficiently large to warrant recognition now
as potantially serious. Continued CFM release at the 1973
level could by the year 2000 produce about half of the
direct climatic effects caused by carbon dioxide increase
over the same period. Continued growth of CFM releases, of
as little as a few percent per year, would lead ultimately,
perhaps in a century or two, to climatic changes of drastic
proportions. The magnitude of both CFM and carbon dioxide
effects, however, are still subject to many uncertainties.

The Impacts

Although the Tukey report deals primarily with
reductions in stratospheric ozone and the "greenhouse"
effect from CFMs, it is clear that the nonfluorinated
halomethanes cause the same phenomena in the troposphere and
stratosphere. The magnitude of the CFM effect was estimated
in terms of the steady-state concentrations that will be
achieved with continued emissions at the 1973 level. These
are estimated to result in ultimate ozone reductions of 7
percent (the central value of the range from 2 to 20
percent), with the "greenhouse" effect being an average
increase in the temperature at the earth's surface of 0.5°C.
The Tukey report carefully assesses the uncertainties in its
predictions of the magnitude of ozone reduction and the
"greenhouse" effect, as well as the consequence or impact of
these phenomena.

It is clear that the present or future effects of the
nonfluorinated halomethanes are or will be substantially
less than those from continued emissions of the CFMs.
Nevertheless, it is useful to consider the nature of the
impact of nonfluorinated halomethanes in the unlikely event
that their emissions to the atmosphere should substantially
increase. Concern over the possible impact from CFMs
indicates the need for limiting and controlling any future
increase in emissions of the nonfluorinated halomethanes.

Nonhuman Biological Effects

For moderate changes in atmospheric ozone content, the
change in uv flux at the surface of the earth involves

radiation only in the biologically harmful 290-320 nm range, the short-wavelength or uv-B regions. The photochemical basis of a number of biological effects resulting from an increase in uv-B radiation is known. These include alterations in proteins and nucleic acids, which result in their failure to perform their biochemical functions properly, to the detriment of the whole organism. For viruses and other microrganisms, damage to the nucleic acid that carries an organism's genetic information can affect both DNA replication and the transcription of genetic information that directs cellular protein synthesis. There are molecular repair processes that mitigate this damage, but these processes are deficient in some cells. Studies on flowering plants and mammals show that the same photochemical damage to DNA occurs in them as in microorganisms. For some species there may be a reserve capacity to deal with greater levels of uv-induced damage, but not for others.

Agricultural Effects

Excessive uv radiation in the vicinity of 300 nm clearly affects plants adversely, as reflected in the reduction of overall growth and photosynthetic activity. However, no readily apparent changes in agricultural production have been associated with changes in mean ozone amounting to perhaps 10 percent over periods of a decade or so. Experimental studies involving both economically important plants and some nonagricultural higher plants have revealed a great heterogeneity of response to supplemented levels of uv radiation. Many species were not sensitive to the levels of uv radiation employed. Sensitive species had inhibited plant growth and development, depressed photosynthetic rates, inhibited pollen germination, and increased somatic mutation rates. Experiments to determine the response of the more sensitive plant species to lesser degrees of ozone depletion (e.g., a 5- or 10-percent ozone reduction) have yet to be simulated. It is apparent, however, that certain types of damage, such as decreased photosynthesis, are cumulative phenomena. Shifts in competitive balance between different plant species might also take place under small increases in uv radiation. Further research would be necessary to determine if such responses occur.

Possible losses in agricultural animal production due to higher levels of uv exposure are also fairly uncertain at this time. While it is known that eye cancer in Hereford cattle, a factor in shortening individual productive life span, is associated with the exposure to solar uv radiation of unpigmented parts of the lid epithelium, other effects of increased levels of radiation in this and other breeds, and in other species of animals, are not known.

Natural Terrestrial and Aquatic Ecosystem Effects

Although the effects of uv radiation on natural ecosystems are difficult to predict, they may be significant. As in agriculture, the individual plant and animal species that make up natural ecosystems differ considerably in their sensitivity to uv radiation. The properties of the system as a whole, however, could be altered if increases in solar uv radiation affect even a few constituent species. One must be concerned not only with direct changes in the populations of a particular organism, but also with other organisms that depend on, or interact with, the sensitive species through some direct or indirect chain of events. In general, only rather small predicted changes in an ecosystem could be viewed with equanimity.

Solar uv-B radiation can penetrate into water; the depth of the penetration depends on the organic and particulate matter content of the water. Two effects are known or can be inferred. The first involves phytoplankton (the microscopic green plants that form the basis of the aquatic food chain), which appear, on the basis of preliminary experiments, to be quite sensitive to uv radiation. The second involves certain fishes and crustaceae, which tend, in their early life stages, to frequent shallow waters where solar uv radiation can penetrate to the bottom. (In some cases, egg clusters actually float on the water surface.) Because of the existence of very active DNA photorepair systems in the cells of marine forms, one may reasonably infer some solar uv radiation damage to DNA.

Effects of Climatic Change on
Agricultural and Natural Systems

Neither the net direction nor the exact magnitude of the change in ground-level climate from CFMs due to ozone depletion and the greenhouse effect can be predicted. Presently, it appears that the principal effect would be due to the direct intensification of the global greenhouse effect. A mean global temperature increase of $0.5°C$ has been used as the possible change resulting from continued release of CFMs at the 1973 rate. This brief discussion of the significance of such a temperature change for biological effects such as crop yields and changes in natural ecosystems is based on two earlier studies: the National Academy of Sciences report by the Climatic Impact Committee (NRC 1975a) and the more comprehensive monograph resulting from the U.S. Department of Transportation Climatic Impact Assessment Program (U.S. DOT 1975).

The model calculations for a $0.5°C$ temperature increase for a variety of major crop species, both for North America and representative areas in other continents, predict changes of usually less than 4 percent of current crop yields. The calculations suggest a slight increase in production for some crops and a slight decrease for others. The net effect on worldwide agricultural productivity is difficult to discern and certainly would be less than the ±4 percent change predicted in particular cases.

For natural systems normally well supplied with water, such as a western coniferous forest and an eastern deciduous forest, the ecosystem models predict that a 0.5°C warming would result in increased primary production of about 5 percent or less. In water-limited ecosystems, such as deserts, the models predict that this amount of warming might result in decreases in primary production of about 4 percent. Although these model predictions indicate relatively small changes on the whole, they do not deal with effects from possible changes in precipitation, cloud cover, circulation patterns, and local changes in climate.

Human Health Effects

Serious effects on humans of exposure to the sun involve the development in some individuals of three major kinds of skin cancers: basal-cell cancers (nonmelanomas); squamous-cell cancers (nonmelanomas); and melanomas. Together, the first two types comprise the most frequently detected cancers in humans and are also the most easily and most successfully treated human cancers. Melanomas are a serious life-threatening hazard and are as common as primary malignant brain tumors.

An increase in melanoma deaths is likely, but not certain, to occur as a consequence of a continuing increase in the rate at which DUV (uv-B in the wavelength range 290-320 nm weighted in accordance with its effectiveness in altering DNA) received at the ground accumulates. The melanoma increase would be delayed while the DUV dose accumulates. A 7-percent ultimate reduction in ozone, with a consequent 14-percent ultimate increase in the DUV accumulation rate, might be expected to cause several hundred melanoma deaths per year. Such ozone and DUV changes would also lead to incidences of basal and squamous-cell carcinomas; these less serious but much more prevalant forms of skin cancer, while rarely causing death, involve significant expense and, occasionally, serious disfigurement.

These detrimental effects on human health would be expected to increase with increased and continuing releases of nonfluorinated halomethanes to the atmosphere, at least for those compounds sufficiently stable to ultimately reach

the stratosphere. It would, therefore, be prudent to assess
continuously and limit substantially increased air emissions
of these compounds.

EXPOSURE AND UPTAKE OF HALOMETHANES BY HUMANS

Halogenated methanes are present in the water we drink,
the air we breathe, and the food we eat. To estimate
potential chronic health effects of these environmental
exposures, it is useful to approximate the amounts of each
halomethane that the human body will take up as a result of
each type of exposure. This section discusses the data base
and the halomethane calculations used to estimate ranges of
human exposures to compounds.

The calculations of human uptake are based on fluid
intake, respiratory volume, and food consumption data for
"reference man," as compiled by the International Commission
for Radiological Protection (ICRP 1975). Table 6.1 presents
some measured fluid intakes for adults and children and
Table 6.2 gives fluid intake data for reference man. Table
6.3 shows the respiratory volumes for reference man based on
a typical breakdown of daily activities. Table 6.4 shows
estimated per capita food consumption of various food groups
by geographical regions, and Table 6.5 summarizes these data
on a worldwide basis and shows the per capita range of
consumption for each food group.

To determine human uptake of halomethanes, the amount of
exposure from each source must also be determined. The
various measurements that different investigators have made
of halomethane concentrations in drinking water, the
atmosphere, and food supplies were described in detail in
Chapter 5 and are summarized in this chapter in an exposure
summary for each source.

Table 6.6 presents the exposure summary of halomethane
concentrations that have been reported in drinking water
through 1976. More recent data reported in an unpublished
EPA study (National Organics Monitoring Study [NOMS], 1977
U.S. EPA, Cincinnati, Ohio), has not been evaluated in this
report. The human uptake of each compound was calculated by
multiplying the minimum, maximum, or median exposure
concentrations by the minimum, maximum, or reference man
fluid intakes. The calculated human uptakes from drinking
water and other ingested fluids under various conditions are
shown in Tables 6.7 through 6.12 for several halomethanes.
These calculations assume that everything included in the
total fluid category contains the indicated halomethane
concentration and that all of the halomethane ingested is
absorbed; commercially produced drinks as well as drinks
reconstituted in the home with municipally treated water are

TABLE 6.1 Measured Fluid Intakes

Subject	Total Fluids[a]		Milk		Tap Water		Waterbased drinks[b]	
	ml/day	liter/yr	ml/day	liter/yr	ml/day	liter/yr	ml/day	liter/yr
Adults (normal conditions)	1,000-2,400	365-876	120-450	44-164	45-730	164-266	320-1,450	117-529
Adults (high environmental temperature to 32°C)	2,840-3,410 3,256 ± 900	1,037-1,245 1,188 ± 329	—	—	—	—	—	—
Adults (moderately active)	3,700	1,351	—	—	—	—	—	—
Children (5-14 yrs)	1,000-1,670	365-610	330-650	120-237	—	—	540-790 ml/day 197-288 l/year	

[a]Numbers in this column are independent measurements; not totals from the following three columns; totals of milk, tap water and waterbased drinks, therefore, do not correspond to total fluid values.
[b]Includes tea, coffee, soft drinks, beer, cider, wine, etc.

SOURCE: ICRP (1975) Report of the Task Group on Reference Man (No. 23). Pergamon Press, Ltd.

TABLE 6.2   Fluid Intake for Reference Man

	Adult Man		Adult Woman		Child (10-yr)	
	ml/day	liter/yr	ml/day	liter/yr	ml/day	liter/yr
Milk	300	109.5	200	73.0	450	164.2
Tap Water	150	54.8	100	36.5	200	73.0
Other	1,500	547.5	1,100	401.5	750	273.8
Total Fluid	1,950	711.8	1,400	511.0	1,400	511.0

SOURCE: ICRP (1975) Report of the Task Group on Reference Man (No. 23). Pergamon Press, Ltd.

TABLE 6.3   Respiratory Volumes for Reference Man
(in liters of air breathed)

	Adult Man	Adult Woman	Child (10 yr)	Infant (1 yr)
8-hr Working Light Activity	9,600	9,100	6,240	2,500 (10-hr)
8-hr Nonoccupational Activity	9,600	9,100	6,240	
8-hr Resting	3,600	2,900	2,300	1,300 (14 hr)
Total (liter/day)	2.3×10^4	2.1×10^4	1.5×10^4	0.38×10^4
Total (liter/year)	8.4×10^6	7.7×10^6	5.5×10^6	1.4×10^6

SOURCE: ICRP (1975) Report of the Task Group on Reference Man (No. 23). Pergamon Press, Ltd.

TABLE 6.4   Per Capita Estimates of World Food Supply Consumption by Region (g/day)

Food Group	Far East	Near East	Africa	Latin America	Europe	North America	Oceania
1. Cereals	404	446	330	281	375	185	243
2. Starchy roots	156	44	473	247	377	136	144
3. Sugar	22	37	29	85	79	113	135
4. Pulses & Nuts	56	47	37	46	15	19	11
5. Vegetables & Fruits	128	398	215	313	316	516	386
6. Meat	24	35	40	102	111	248	312
7. Eggs	3	5	4	11	23	55	31
8. Fish	27	12	16	18	38	26	22
9. Milk	51	214	96	240	494	850	574
10. Fats and Oils	9	20	19	24	44	56	45

1. In terms of flour and milled rice.
2. Includes sweet potatoes, cassava, and other edible roots.
3. Includes raw sugar; excludes syrups and honey.
4. Includes cocoa-beans.
5. In terms of fresh equivalent.
6. Includes offal, poultry, and game expressed in terms of carcass weight, excluding slaughter fats.
7. Fresh egg equivalent.
8. Landed weight.
9. Excludes butter; includes mild products as fresh milk, equivalent.
10. Pure fat content.

SOURCE: ICRP (1975) Report of the Task Group on Reference Man (No. 23). Pergamon Press, Ltd.

TABLE 6.5 Summary of Per Capita Estimates of World Food Supply Consumption

Food Groups	Worldwide Min (g/day)	Max (g/day)	Min (kg/yr)	Max (kg/yr)
1. Cereals	185	446	67.5	162.8
2. Starchy Roots	44	473	16.1	172.6
3. Sugar	22	135	8.0	49.3
4. Pulses & Nuts	11	56	4.0	20.4
5. Vegetables & Fruits	128	516	46.7	188.3
6. Meats	24	312	8.8	113.9
7. Eggs	3	55	1.1	20.1
8. Fish	12	38	4.4	13.9
9. Milk	51	850	18.6	310.2
10. Fats & Oils	9	56	3.3	20.4

1. In terms of flour and milled rice.
2. Includes sweet potatoes, cassava, and other edible roots.
3. Includes raw sugar; excludes syrups and honey.
4. Includes cocoa-beans.
5. In terms of fresh equivalent.
6. Includes offal, poultry, and game expressed in terms of carcass weight, excluding slaughter fats.
7. Fresh egg equivalent.
8. Landed weight.
9. Excludes butter; includes mild products as fresh milk, equivalent.
10. Pure fat content.

SOURCE: ICRP (1975) Report of the Task Group on Reference Man (No. 23). Pergamon Press, Ltd.

TABLE 6.6 Environmental Exposure Summary of Halomethane Concentrations in Drinking Water ($\mu g/l$), ppb

Compound	Range	Reference
Chloroform	1.7-4.4 (well water source)	Bellar et al. (1974)
	37.3-152 (surface water source)	Bellar et al. (1974)
	<0.1-311 (median = 21, finished water)	Symons et al. (1975)
	<0.1-366 (finished water)	Symons (1976)
Carbon Tetrachloride	<2.0-3.0	Symons et al. (1975)
Bromoform	<0.8-92	"
Bromodichloromethane	0.3-116 (median = 6)	"
Dibromochloromethane	0.4-110 (median = 12)	"
Methylene Chloride	Mean-1.0[a]	U.S. EPA (1975)
	Maximum-7.0[a]	"

[a]These are values for only those 8 percent of the total number of communities studied (83) which had positive results.

TABLE 6.7 Chloroform Uptake from Fluids (mg/yr) Assuming 100 Percent Absorption[a]

Exposure	Fluid	Adult Man			Adult Woman			Child (5-14 yr)			(10 yr)
		Min. Fluid Intake	Max. Fluid Intake	Refer. Man Intake	Min. Fluid Intake	Max. Fluid Intake	Refer. Man Intake	Min. Fluid Intake	Max. Fluid Intake	Refer. Man Intake	
Minimum Concentration Exposure (0.0001 mg/l)	Tap Water	0.002	0.027	0.005	0.002	0.027	0.004			0.007	
	Other[b]	0.012	0.053	0.055	0.012	0.053	0.040	0.020	0.029	0.027	
	Total Fluid	0.037	0.135	0.071	0.037	0.135	0.051	0.036	0.061	0.051	
Median Concentration Exposure (0.021 mg/l)	Tap Water	0.34	5.59	1.15	0.34	5.59	0.77			1.53	
	Other[b]	2.45	11.1	11.5	2.45	11.1	8.43	4.14	6.06	5.75	
	Total Fluid	7.67	28.4	14.9	7.67	28.4	10.7	7.67	12.8	10.7	
Maximum Concentration Exposure (0.366 mg/l)	Tap Water	6.01	97.5	20.1	6.01	97.5	13.4			26.7	
	Other[b]	42.7	194	200	42.7	194	147	72.1	106	100	
	Total Fluid	134	494	261	134	494	187	134	223	187	

[a] Calculated by multiplying the exposure concentration (μg/l) (from Table 6.6) x fluid intake (l/yr) (from Table 6.1 for minimum and maximum intakes and Table 6.2 for reference man intakes), and dividing by 1000 μg/mg = mg/yr.
[b] Includes water-based drinks, such as tea, coffee, soft drinks, beer, cider, wine.

TABLE 6.8 Carbon Tetrachloride Uptake from Fluids (mg/yr) Assuming 100 Percent Absorption[a]

Exposure	Fluid	Adult Man			Adult Woman			Child		
								(5-14 yr)		(10 yr)
		Min. Fluid Intake	Max. Fluid Intake	Refer. Man Intake	Min. Fluid Intake	Max. Fluid Intake	Refer. Man Intake	Min. Fluid Intake	Max. Fluid Intake	Refer. Man Intake
Minimum Concentration Exposure (0.002 mg/l)	Tap Water	0.03	0.53	0.11	0.03	0.53	0.07	0.39	0.58	0.15
	Other[b]	0.23	1.06	1.10	0.23	1.06	0.80			0.55
	Total Fluid	0.73	2.70	1.42	0.73	2.70	1.02	0.73	1.22	1.02
Maximum Concentration Exposure (0.003 mg/l)	Tap Water	0.05	0.80	0.16	0.05	0.80	0.11	0.59	0.86	0.22
	Other[b]	0.35	1.59	1.64	0.35	1.59	1.20			0.82
	Total Fluid	1.10	4.05	2.14	1.10	4.05	1.53	1.10	1.83	1.53

[a] Calculated by multiplying the exposure concentration (μg/l) (from Table 6.6) x fluid intake (l/yr) (from Table 6.1 for minimum and maximum intakes and Table 6.2 for reference man intakes), and dividing by 1000 μg/mg = mg/yr.
[b] Includes water-based drinks, such as tea, coffee, soft drinks, beer, cider, wine.

TABLE 6.9 Bromoform Uptake from Fluids (mg/yr) Assuming 100 Percent Absorption[a]

Exposure	Fluid	Adult Man			Adult Woman			Child (5-14 yr)		(10 yr)
		Min. Fluid Intake	Max. Fluid Intake	Refer. Man Intake	Min. Fluid Intake	Max. Fluid Intake	Refer. Man Intake	Min. Fluid Intake	Max. Fluid Intake	Refer. Man Intake
Minimum Concentration Exposure (0.0008 mg/l)	Tap Water	0.01	0.21	0.04	0.01	0.21	0.03	0.16	0.23	0.06
	Other[b]	0.09	0.42	0.44	0.09	0.42	0.32	0.29	0.49	0.22
	Total Fluid	0.29	1.08	0.57	0.29	1.08	0.41			0.41
Maximum Concentration Exposure (0.092 mg/l)	Tap Water	1.51	24.5	5.04	1.51	24.5	3.36	18.1	26.5	6.72
	Other[b]	10.7	48.7	50.4	10.7	48.7	36.9	33.6	56.1	25.2
	Total Fluid	33.6	124	65.5	33.6	124	47.0			47.0

[a] Calculated by multiplying the exposure concentration (μg/l) (from Table 6.6) x fluid intake (l/yr) (from Table 6.1 for minimum and maximum intakes and Table 6.2 for reference man intakes), and dividing by 1000 μg/mg = mg/yr.
[b] Includes water-based drinks, such as tea, coffee, soft drinks, beer, cider, wine.

TABLE 6.10 Bromodichloromethane Uptake from Fluids (mg/yr) Assuming 100 Percent Absorption[a]

Exposure	Fluid	Adult Man			Adult Woman			Child (5-14 yr)		(10 yr)
		Min. Fluid Intake	Max. Fluid Intake	Refer. Man Intake	Min. Fluid Intake	Max. Fluid Intake	Refer. Man Intake	Min. Fluid Intake	Max. Fluid Intake	Refer. Man Intake
Minimum Concentration Exposure (0.0003 mg/l)	Tap Water	0.005	0.08	0.02	0.005	0.08	0.01			0.02
	Other[b]	0.04	0.16	0.16	0.04	0.16	0.12	0.06	0.09	0.08
	Total Fluid	0.11	0.41	0.21	0.11	0.41	0.15	0.11	0.18	0.15
Median Concentration Exposure (0.006 mg/l)	Tap Water	0.10	1.60	0.33	0.10	1.60	0.22			0.44
	Other[b]	0.70	3.18	3.28	0.70	3.18	2.41	1.18	1.73	1.64
	Total Fluid	2.19	8.10	4.27	2.19	8.10	3.07	2.19	3.66	3.07
Maximum Concentration Exposure (0.116 mg/l)	Tap Water	1.90	30.9	6.36	1.90	30.9	4.23			8.47
	Other[b]	13.5	61.4	63.5	13.5	61.4	46.6	22.9	33.4	31.8
	Total Fluid	42.3	157	82.6	42.3	157	59.3	42.3	70.7	59.3

[a] Calculated by multiplying the exposure concentration (μg/l) (from Table 6.6) x fluid intake (l/yr) (from Table 6.1 for minimum and maximum intakes and Table 6.2 for reference man intakes), and dividing by 1000 μg/mg = mg/yr.
[b] Includes water-based drinks, such as tea, coffee, soft drinks, beer, cider, wine.

TABLE 6.11 Dibromochloromethane Uptake from Fluids (mg/yr) Assuming 100 Percent Absorption[a]

Exposure	Fluid	Adult Man			Adult Woman			Child (5-14 yr)		(10 yr)
		Min. Fluid Intake	Max. Fluid Intake	Refer. Man Intake	Min. Fluid Intake	Max. Fluid Intake	Refer. Man Intake	Min. Fluid Intake	Max. Fluid Intake	Refer. Man Intake
Minimum Concentration Exposure (0.0004 mg/l)	Tap Water	0.007	0.11	0.02	0.007	0.11	0.01	0.08	0.12	0.03
	Other[b]	0.05	0.21	0.22	0.05	0.21	0.16			0.11
	Total Fluid	0.15	0.54	0.28	0.15	0.54	0.20	0.15	0.24	0.20
Median Concentration Exposure (0.001 mg/l)	Tap Water	0.02	0.27	0.05	0.02	0.27	0.04	0.20	0.29	0.07
	Other[b]	0.12	0.53	0.55	0.12	0.53	0.40			0.27
	Total Fluid	0.36	1.35	0.71	0.36	1.35	0.51	0.36	0.61	0.51
Maximum Concentration Exposure (0.110 mg/l)	Tap Water	1.81	29.3	6.03	1.81	29.3	4.02	21.7	31.7	8.03
	Other[b]	12.8	58.2	60.2	12.8	58.2	44.2			30.2
	Total Fluid	40.2	149	78.3	40.2	149	56.2	40.2	67.1	56.2

[a] Calculated by multiplying the exposure concentration (μg/l) (from Table 6.6) x fluid intake (1/yr) (from Table 6.1 for minimum and maximum intakes and Table 6.2 for reference man intakes), and dividing by 1000 μg/mg = mg/yr.
[b] Includes water-based drinks, such as tea, coffee, soft drinks, beer, cider, wine.

TABLE 6.12 Methylene Chloride Uptake from Fluids (mg/yr) Assuming 100 Percent Absorption[a]

Exposure	Fluid	Adult Man			Adult Woman			Child (5-14 yr)			(10 yr)
		Min. Fluid Intake	Max. Fluid Intake	Refer. Man Intake	Min. Fluid Intake	Max. Fluid Intake	Refer. Man Intake	Min. Fluid Intake	Max. Fluid Intake	Refer. Man Intake	Refer. Man Intake
Mean Concentration Exposure[c] (0.001 mg/l)	Tap Water	0.02	0.27	0.05	0.02	0.27	0.04				0.07
	Other[b]	0.12	0.53	0.55	0.12	0.53	0.40	0.20	0.29		0.27
	Total Fluid	0.36	1.35	0.71	0.36	1.35	0.51	0.36	0.61		0.51
Maximum Concentration Exposure[c] (0.007 mg/l)	Tap Water	0.11	1.86	0.38	0.11	1.86	0.26				0.51
	Other[b]	0.82	3.70	3.83	0.82	3.70	2.81	1.38	2.02		1.92
	Total Fluid	2.56	9.45	4.98	2.56	9.45	3.58	2.56	4.27		3.58

[a] Calculated by multiplying the exposure concentration (μg/l) (from Table 6.6) x fluid intake (l/yr) (from Table 6.1 for minimum and maximum intakes and Table 6.2 for reference man intakes), and dividing by 1000 μg/mg = mg/yr.
[b] Includes water-based drinks, such as tea, coffee, soft drinks, beer, cider, wine.
[c] These are values for only those 8 percent with positive values of a sample of 83 communities (Symons 1976).

likely to be contaminated. Some drinks that are not as directly water-based, such as milk, are included in the total fluid category making the uptake calculations more conservative, from the human health perspective, than the total of the tap water and the "other" categories.

Table 6.13 is an exposure summary of halomethane concentrations that have been reported in the ambient air. The human uptake of each compound was calculated by multiplying the minimum, typical, or maximum environmental concentrations by the respiratory volumes for each category of reference man (Table 6.3) and is shown in Tables 6.14 through 6.17. The tables for methyl iodide and methyl chloride assume 100 percent absorption of the amount inhaled. The tables showing the uptake of chloroform and carbon tetrachloride also state the range, by percent, of each of these compounds that is absorbed, and not exhaled, by humans. The minimum and maximum amounts absorbed were calculated by multiplying the range of absorptions by the amounts inhaled. In the steady state, the percentage absorbed by the lungs may be considerably less than the ranges used. Nevertheless, in the absence of steady-state data, it is conservative to assume and use these possibly higher rates of absorption.

The environmental exposure summary of concentrations of carbon tetrachloride and chloroform that have been reported in various food groups is shown in Table 6.18. In Table 6.19, the human uptake of chloroform and carbon tetrachloride from each food group was calculated by assuming 100 percent absorption of the amount ingested and multiplying the minimum or maximum exposure concentration by the minimum or maximum per capita food consumption shown in Table 6.5. The total uptake from all food supplies was then calculated for various exposure and food intake conditions.

Relative Source Strengths

Only chloroform and carbon tetrachloride have human uptake data available for all three environmental sources of halomethanes--fluid intake, atmosphere, and food supplies-- and it is therefore possible to compare the uptake from each of these sources only for these two compounds. The relative strengths of these compounds as environmental sources are shown in Table 6.20 and represented graphically in Figure 6.1.

At minimum levels of exposure, the uptake of carbon tetrachloride from the atmosphere is about five times greater than it is from fluid intake, and at maximum exposure levels, its uptake from the atmosphere can be up to 150 times greater than from fluids. At typical exposure

TABLE 6.13 Environmental Exposure Summary of Halomethane Concentrations in the Outdoor Atmosphere (ppb, by volume)

	Chloroform	Carbon Tetrachloride	Methyl Iodide	Methyl Chloride
Minimum	0.02	0.12	0.001	0.55
Typical	0.20	0.15	0.02	0.75
Maximum	15.0*	18.0*	3.8*	1.50

*These values were measured in Bayonne, N.J. (Lillian et al. 1975) and are unusually high, presumably because of high industrial input.

SOURCE: Derived from Table 5.16.

TABLE 6.14 Chloroform Uptake from the Outdoor Atmosphere (mg/yr)[a]

	Adult Man			Adult Woman		
	Minimum Exposure	Typical Exposure	Maximum Exposure	Minimum Exposure	Typical Exposure	Maximum Exposure
Average Amount Inhaled (mg/yr)[b]	0.82	8.2	615	0.75	7.5	564
Min. Absorbed (mg/yr)[c]	0.41	4.1	304	0.37	3.7	279
Max. Absorbed (mg/yr)[c]	0.63	6.3	474	0.58	5.8	434

	Child (10-yr)			Infant (1-yr)		
	Minimum Exposure	Typical Exposure	Maximum Exposure	Minimum Exposure	Typical Exposure	Maximum Exposure
Average Amount Inhaled (mg/yr)[b]	0.54	5.4	403	0.14	1.4	102
Min. Absorbed (mg/yr)[c]	0.27	2.7	199	.07	0.7	50.4
Max. Absorbed (mg/yr)[c]	0.41	4.1	310	0.11	1.1	78.5

[a] Conversion factor, ppb chloroform, by volume, to mg/l = 4.88 x 10^{-6}. Percent absorption range: 49.4%-77%, calculated from the concentration of chloroform in the inhaled air and that found in the exhaled breath (Lehmann and Hasegawa 1910).
[b] Calculated by applying respiratory volume per year (Table 6.3) to the general range of atmospheric concentrations (Table 6.13), using the stated conversion factor.
[c] Calculated by applying the range of percent absorptions to the total amount inhaled per year.

TABLE 6.15 Carbon Tetrachloride Uptake from the Outdoor Atmosphere (mg/yr)[a]

	Adult Man			Adult Woman		
	Minimum Exposure	Typical Exposure	Maximum Exposure	Minimum Exposure	Typical Exposure	Maximum Exposure
Average Amount Inhaled (mg/yr)[b]	6.3	7.9	951	5.8	7.3	872
Min. Absorbed (mg/yr)[c]	3.6	4.5	542	3.3	4.1	457
Max. Absorbed (mg/yr)[c]	4.1	5.2	618	3.8	4.7	567

	Child (10-yr)			Infant (1-yr)		
	Minimum Exposure	Typical Exposure	Maximum Exposure	Minimum Exposure	Typical Exposure	Maximum Exposure
Average Amount Inhaled (mg/yr)[b]	4.2	5.2	623	1.1	1.3	159
Min. Absorbed (mg/yr)[c]	2.4	3.0	355	0.6	0.6	90.0
Max. Absorbed (mg/yr)[c]	2.7	3.4	405	0.7	0.9	103

[a] Conversion factor, ppb carbon tetrachloride, by volume, to mg/l = 6.29 x 10^{-6}. Percent absorption range: 57%-65%, calculated from the concentration of carbon tetrachloride in the inhaled air and that found in the exhaled breath (Lehmann and Schmidt-Kehl 1936).
[b] Calculated by applying respiratory volume per year (Table 6.3) to the general range of atmospheric concentrations (Table 6.13), using the stated conversion factor.
[c] Calculated by applying the range of percent absorptions to the total amount inhaled per year.

TABLE 6.16 Methyl Iodide Uptake from the Atmosphere (mg/yr) Assuming 100 Percent Absorption[a]

	Adult Man			Adult Woman		
	Minimum Exposure	Typical Exposure	Maximum Exposure	Minimum Exposure	Typical Exposure	Maximum Exposure
Average Amount Inhaled (mg/yr)[b]	0.04	0.85	162	0.04	0.78	149
	Child (10-yr)			Infant (1-yr)		
	Minimum Exposure	Typical Exposure	Maximum Exposure	Minimum Exposure	Typical Exposure	Maximum Exposure
Average Amount Inhaled (mg/yr)[b]	0.03	0.56	106	0.06	0.14	27.0

[a] Conversion factor, ppb methyl iodide, by volume, to mg/l = 5.08×10^{-6}.
[b] Calculated by applying respiratory volume per year (Table 6.3) to the range of atmospheric concentrations (Table 6.13), using the stated conversion factor.

TABLE 6.17 Methyl Chloride Uptake from the Atmosphere (mg/yr) Assuming 100 Percent Absorption[a]

	Adult Man			Adult Woman		
	Minimum Exposure	Typical Exposure	Maximum Exposure	Minimum Exposure	Typical Exposure	Maximum Exposure
Average Amount Inhaled (mg/yr)[b]	9.5	13	26	8.7	12	24
	Child (10-yr)			Infant (1-yr)		
	Minimum Exposure	Typical Exposure	Maximum Exposure	Minimum Exposure	Typical Exposure	Maximum Exposure
Average Amount Inhaled (mg/yr)[b]	6.2	8.5	17	1.6	2.2	4.3

[a] Conversion factor, ppb methyl chloride, by volume, to mg/l = 2.065 × 10^{-6}.
[b] Calculated by applying respiratory volume per year (Table 6.3) to the range of atmospheric concentrations (Table 6.13), using the stated conversion factor.

TABLE 6.18 Exposure Summary of Halomethane Concentrations in Food Supplies (ppb by weight, µg/kg)

	Carbon Tetrachloride		Chloroform	
Food Group	Min.	Max.	Min.	Max.
Milk Products	0.2	14.0	5.0	33.0
Eggs (average)		0.5		1.4
Meats	7.0	9.0	1.0	4.0
Fats and Oils	0.7	18.0	2.0	10.0
Vegetables and Fruits	3.0	8.0	2.0	18.0
Fish and Seafood	0.1	6.0	3.0	150.0

SOURCE: Adapted from McConnell et al. (1975) and Pearson and McConnell (1975).

TABLE 6.19 Halomethane Uptake from Food Supplies (mg/yr) Assuming 100 Percent Absorption[a]

Food Group[b]	Worldwide Food Intake[c]	Carbon Tetrachloride		Chloroform	
		Min. Exposure	Max. Exposure	Min. Exposure	Max. Exposure
Milk Products	Min.	0.004	0.26	0.09	0.61
	Max.	0.06	4.34	1.55	10.2
Eggs (average)	Min.	0.001		0.002	
	Max.	0.01		0.03	
Meats	Min.	0.06	0.08	0.009	0.04
	Max.	0.80	1.02	0.11	0.46
Fats and Oils	Min.	0.002	0.06	0.007	0.03
	Max.	0.01	0.37	0.04	0.20
Vegetables and Fruits	Min.	0.14	0.37	0.09	0.84
	Max.	0.56	1.51	0.38	3.39
Fish and Seafood	Min.	<0.001	0.03	0.01	0.66
	Max.	0.001	0.08	0.04	2.09

Total: Carbon Tetrachloride Uptake From All Food Supplies (mg/yr)
  Minimum Exposure-Minimum Intake   0.21
  Average Exposure and Intake   1.12
  Maximum Exposure-Maximum Intake   7.33

Total: Chloroform Uptake from All Food Supplies (mg/yr)
  Minimum Exposure-Minimum Intake   0.21
  Average Exposure and Intake   2.17
  Maximum Exposure-Maximum Intake   16.37

[a]Calculated by applying worldwide food intake ranges for various food groups (Table 6.5) to the range of concentrations found in various foods (Table 6.18).
[b]As used in Tables 6.4 and 6.5.
[c]From Table 6.5.

TABLE 6.20 Relative Human Uptake of Carbon Tetrachloride (CCl_4) and Chloroform ($CHCl_3$) from Environmental Sources (mg/year)

	At Minimum Exposure Levels[a]					
	Adult Man		Adult Woman		Child	
Source	CCl_4	$CHCl_3$	CCl_4	$CHCl_3$	CCl_4	$CHCl_3$
Fluid Intake	0.73	0.037	0.73	0.037	0.73	0.036
Atmosphere	3.60	0.41	3.30	0.37	2.40	0.27
Food Supply	0.21	0.21	0.21	0.21	0.21	0.21
Total	4.54	0.66	4.24	0.62	3.34	0.52

	At Typical Exposure Levels[b]					
	Adult Man		Adult Woman		Child	
Source	CCl_4	$CHCl_3$	CCl_4	$CHCl_3$	CCl_4	$CHCl_3$
Fluid Intake	1.78	14.90	1.28	10.70	1.28	10.70
Atmosphere	4.80	5.20	4.40	4.70	3.20	3.40
Food Supply	1.12	2.17	1.12	2.17	1.12	2.17
Total	7.70	22.27	6.80	17.57	5.60	16.27

	At Maximum Exposure Levels[c]					
	Adult Man		Adult Woman		Child	
Source	CCl_4	$CHCl_3$	CCl_4	$CHCl_3$	CCl_4	$CHCl_3$
Fluid Intake	4.05	494	4.05	494	1.83	223
Atmosphere	618	474	567	434	405	310
Food Supply	7.33	16.4	7.33	16.4	7.33	16.4
Total	629	984	578	944	414	549

[a]Minimum conditions of all variables assumed: Minimum exposure-minimum intake for fluids; minimum exposure-minimum absorption for atmosphere; and minimum exposure-minimum intake for food supplies.
[b]Typical conditions of all variables assumed.
For CCl_4: 0.0025 mg/l-reference man intake for fluids; average of typical minimum and maximum absorption for atmosphere; and average exposure and intake for food supplies.
For $CHCl_3$: median exposure-reference man intake for fluids; average of typical minimum and maximum absorption for atmosphere; and average exposure and intake for food supplies.
[c]Maximum conditions of all variables assumed: maximum exposure-maximum intake for fluids; maximum exposure-maximum absorption for atmosphere; and maximum exposure-maximum intake for food supplies.

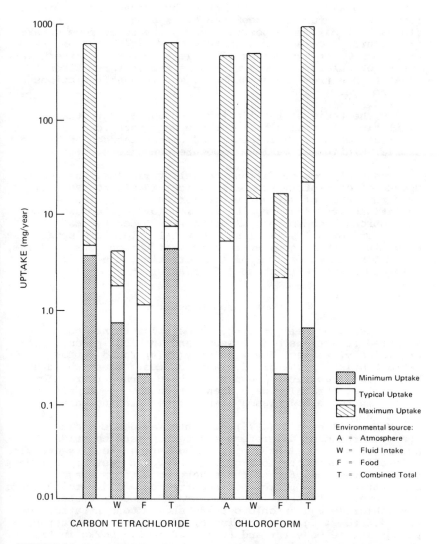

FIGURE 6.1 Relative uptake of carbon tetrachloride and chloroform by adult man from fluid intake, atmosphere, and food supply (mg/year).

levels, the uptake from the atmosphere is around three times greater than from fluids.

The uptake of chloroform from the atmosphere at minimum levels of exposure is about ten times greater than from fluids. At maximum exposure levels, the chloroform uptake from fluids becomes approximately equal to that from the atmosphere. At typical exposure levels, however, the human uptake of chloroform from fluids is two to three times greater than from the atmosphere, with slight variations by sex and age.

On the basis of the limited food supply data that are available, it appears that for the most part food is the smallest source of chloroform and carbon tetrachloride under similar minimum or maximum exposure conditions.

Under minimum conditions for all variables, the total uptake of chloroform from all sources is six to seven times less than that of carbon tetrachloride. As typical conditions are attained, the total chloroform uptake from all sources becomes more significant. Under typical conditions of all variables, the total chloroform uptake is about three times greater than that of carbon tetrachloride. Under maximum conditions of all variables, the total chloroform uptake is about 1.5 times greater.

Confirming Exposures

To confirm their presence, some halomethanes have been measured in various body tissues and in blood plasma. In preliminary studies, Dowty et al. (1974) found carbon tetrachloride in the blood plasma of test groups who drank New Orleans tap water known to contain carbon tetrachloride. The carbon tetrachloride concentration was substantially higher in the blood plasma than in the drinking water.

McConnell et al. (1975) presented evidence indicating the occurrence of chloroform and carbon tetrachloride in human tissues of eight subjects who were 48 to 82 years old. Chloroform concentrations of 5 to 68 µg/kg (wet tissue) were found in the body fat of the eight subjects (average concentration, 51 µg/kg). Concentrations of 1.0 to 10 µg/kg were found in kidney, liver and brain tissue. Carbon tetrachloride and trichloroethane showed concentrations of from 1.6 to 24 µg/kg (wet tissue) (average concentration, 8 µg/kg) in the body fat and from <1.0 to 5.1 µg/kg in kidney, liver, and brain tissue.

Methyl Chloride Exposure from Cigarette Smoke

Human exposure to methyl chloride from cigarette smoke is local in nature and affects discrete target populations. Several estimates of methyl chloride production from cigarette smoking have been made. Corn (1974) reported a method for calculating concentrations of various pollutants, including methyl chloride, from tobacco combustion. The equation takes into account factors affecting concentration, such as room volume, initial pollutant concentration, ventilation, pollutant concentrations in the ventilation air, mixing time, and production rate of the pollutant. Corn also presented sample calculations of the equilibrium concentration of methyl chloride in an experiment performed by Owens and Rossano (1969). This concentration was 0.1 mg/m^3, based on calculations for a 30 x 30 x 10-foot room with 30 persons each smoking 1.29 cigarettes per hour and air circulation rate of 30 cu ft/min/person, including 5 cu ft/min/person of outdoor air. The net methyl chloride production rate was 0.87 mg/hr/person or about 0.674 mg/cigarette, producing a concentration of 0.1 mg/m^3 that is equal to about 49,500 ppt (25°C, 1 atm).

Chopra and Sherman (1972) reported a methyl chloride production rate of about 2.0 mg per cigarette smoked (assuming 1 cigarette = ~1 gm tobacco). Smoking one cigarette in a typical 8 x 10 x 8-foot room without ventilation would produce a methyl chloride concentration of about 60,000 ppt (25°C, 1 atm) at this production rate.

Philippe and Hobbs (1956) reported a methyl chloride production rate of 3.2 x 10^{-2} ml per puff from cigarette smoking. Based on 8 to 13 puffs per cigarette, the methyl chloride production rate would be about 0.518 to 0.841 mg per cigarette smoked (25°C, 1 atm). In a typical 8 x 10 x 8-foot room without ventilation, smoking one cigarette would produce a methyl chloride concentration of about 15,000 to 24,000 ppt (25°C, 1 atm) at this production rate.

Based on these methyl chloride production rates, the range of methyl chloride concentrations in occupied spaces is roughly 15,000 to 60,000 ppt (25°C, 1 atm). This range is reasonably consistent with the indoor methyl chloride concentration of 20,000+ ppt measured by Rasmussen (Table 5.17) in an apartment following cigarette smoking. It appears that the calculated methyl chloride concentration in a typical 8 x 10 x 8-foot room with no ventilation, based on the production rate determined in each study, would be within this range of concentrations:

Production Rate (mg CH_3Cl/cigarette)	Calculated CH_3Cl Concentration (ppt) in typical room space	Reference
0.518	15,000	Philippe and Hobbs (1956)
0.674	19,000	Owens and Rossano (1969)
0.841	24,000	Philippe and Hobbs (1956)
2.00	60,000	Chopra and Sherman (1972)

ESTIMATE OF HUMAN HEALTH EFFECTS AT ACTUAL EXPOSURE LEVELS

Human risk assessment is possibly the most difficult problem associated with environmental decision making. It requires the use of epidemiological and laboratory findings for the quantitative prediction of effects of environmental agents on human health. The need for good quantitative prediction is clear in order to evaluate the economics of environmental controls.

The health effects of concern in this report exclude those from the high exposures that may be found in the industrial workplace. An assessment of exposures of a general environmental nature requires the prediction of effects at low exposures of lifetime duration. Most attempts to link various environmental pollution levels with vital statistics data are inconclusive because of the difficulty in detecting a small increase in some health effect, plus the confusion caused by natural variations in environmental quality and demographic characteristics. This is generally true at this time for the nonfluorinated halomethanes.

Health effects at environmental levels therefore must be estimated from data obtained at much higher exposure levels. For adverse biological effects other than mutagenesis and carcinogenesis, it is generally felt that there are thresholds operating. Upon examination of the data and the use of safety factors, permissible occupational levels are determined. This has in fact been done recently by the National Institute of Occupational Safety and Health (NIOSH) (U.S. DHEW 1974, 1975, 1976) for carbon tetrachloride, methylene chloride, and chloroform (2 ppm, 75 ppm, and 10 ppm, respectively, for time-weighted average exposure in air, over a 40-hour work week). Such permissible occupational levels can give an indication of concern when environmental levels begin to approach them. For perspective, in one preliminary study (Table 5.14) the highest indoor ambient air concentrations were about 0.4, 23, and 0.7 ppm for carbon tetrachloride, methylene chloride, and chloroform respectively.

However, the heterogeneity of the general population, with its various susceptible subgroups, must be considered. Further, there is little understanding of the combined effects of many environmental agents when each is at a relatively low level. The problem of possible synergistic effects is particularly troublesome. For example, the NIOSH permissible level for methylene chloride depends upon carbon monoxide levels because the toxicities of the two compounds are felt to be additive. Even ignoring such complications, there does not appear to be a satisfactory method for extrapolating general toxicity information to predict possible health effects at low exposure levels. Probably, the best that can be done is to direct attention to those agents that begin to approach environmental levels that are considered limits for occupational exposures.

Dealing with carcinogenesis is somewhat different for the reason that the carcinogenic process is generally irreversible and self-replicating. Further, because it is thought that cancer often originates from a single somatic cell mutation, assumptions of thresholds may be invalid. There has been considerable recent activity in the study of various models for carcinogenesis and their mathematical application to low dose estimation (Crump et al. 1976, Hoel et al. 1975). It is hoped that use of these methods will permit fairly close estimation of low-dose risks. However, these methods do not take into account changes in metabolic pathways at various dose levels, synergism, or susceptible subgroups. How great a problem these difficulties create and what to do about them is simply not known.

General Toxicity

For the specific compounds under consideration, there is only a limited amount of information available on human toxicity. While most halogenated methanes are known to be moderately or even highly toxic on an acute exposure basis, little if any data exist on the minimum acute toxic dosages for most instances of inhalation or ingestion. As a group, these compounds generally have narcotic properties at high dose levels and may cause injury to the central nervous system, liver, and kidneys.

All available information on the carcinogenicity, mutagenicity, and/or teratogenicity of the halogenated methanes is based on animal studies. Table 6.21 presents a toxicity summary of the compounds in this study. Appendix B presents detailed information on acute and chronic toxicological effects in man and animals. Three compounds (carbon tetrachloride, chloroform, and methyl iodide) have been found to be carcinogenic in at least one strain of experimental animals (see Appendix B for levels). The others have not been adequately tested but should be considered as possibly suspect due to their alkylating properties.

Methyl chloride, methyl bromide, and bromoform have all been found to test positive in the Ames system, while methylene chloride was found not mutagenic in Drosophila. In addition, carbon tetrachloride has been identified as a potential chromosome-breaking agent. None of the other compounds under consideration appear to have been sufficiently tested to assess their mutagenic potential.

Data on teratogenic potential and feto-toxicity are not available for most of the compounds. Based on the limited results that have appeared in the literature, methylene chloride has displayed some indication of possible feto-toxicity; carbon tetrachloride does not seem to produce terata; and chloroform is feto-toxic, and not teratogenic (Appendix B).

Extrapolation and Carcinogenic-Risk Estimation

The process of estimating human cancer risk from animal-based experiments is fraught with a variety of complex issues that are often difficult or impossible to totally resolve. For example, the appropriateness of any given animal species for assessing the carcinogenic potential of a particular chemical agent in man is obviously dependent on species differences in absorption or uptake, distribution and storage, metabolism, and excretion. Furthermore, animal experiments are generally conducted on highly inbred, homogeneous strains, while the human population is genetically heterogeneous. In many instances, the subgroup at highest risk and, therefore, of greatest interest may represent only a small portion of the exposed population. Finally, laboratory experiments are generally focused on the activity of a single compound, while in the human environment man is exposed simultaneously to a complex variety of potentially hazardous agents and dietary factors whose joint action may be greater than the simple addition of individual effects might indicate. The net effect of these differences between a segment of the human population and animal surrogates is that man may be more or less sensitive than the experimental animals used for carcinogenesis screening studies.

Most mathematical models that have been developed to estimate the risk from long-term exposure to a potential carcinogen are concerned only with the problem of low-dose extrapolation. That is, animal data from experiments conducted at dose levels high enough to produce tumors in an appreciable percentage of the test animals are used to estimate the corresponding risk at ambient levels for the human population. The only consideration generally given to the problems with animal-to-man extrapolation is to express the data on the basis of dose per unit of surface area. A variety of factors affecting the action of the presumed carcinogen (e.g., distribution and persistence, cell-division rates, and even total lifespan) are related to body size, and use of this dose metric (i.e., dose per unit of surface area), appears to reduce calculated species differences in many instances.

The magnitude of the risk estimate for any given low-dose level will obviously depend heavily on the specific form that is assumed for the dose-response curve in making the estimate. Models that are customarily proposed include the threshold model, the multi-stage model, and the probit-type models (Hoel et al. 1975).

The threshold model assumes some critical level of exposure below which the carcinogenic process will not be initiated. The arguments in support of no-effect levels generally center around the concept of proper detoxification and adequate repair systems including DNA repair and immunosuppression. The reason for high dose effects is then attributed to system overloads and thus a breakdown of the body's natural defense system. Also, there is the possibility that activation of the system which leads to the carcinogenic products will not be initiated at low dose levels.

The opposing views argue that since there are no known chemical carcinogens that can produce tumors that are not found to occur in the absence of that chemical, the threshold hypothesis requires the assumption that the carcinogen in question acts by some novel mechanism (i.e., independent of all ongoing processes) on the target organ or site. If this were to occur, then it might be possible for the cell repair process to impede the induced cell transformation below a certain (threshold) level of exposure. However, if there is a non-zero probability that the repair function will not totally inhibit transformation, then this theory cannot be substantiated. Furthermore, since the human population is genetically quite heterogenous, separate thresholds would have to be

TABLE 6.21 Toxicity Summary, Nonfluorinated Halomethanes

Characteristics	Methyl Chloride (CH_3Cl)	Methyl Bromide (CH_3Br)	Methyl Iodide (CH_3I)	Methylene Chloride (CH_2Cl_2)	Methylene Bromide (CH_2Br_2)	Methylene Iodide (CH_2I_2)
TLV-TWA:[a]	100 ppm	15 ppm (skin)	5 ppm (skin)	75 ppm	none	none
Acute Toxicity: Minimum Acute Toxic Doses	moderately toxic	highly toxic	highly toxic	mildly toxic	unknown	unknown
Inhalation	>500 ppm (mild symptoms)	50-500 ppm (mild symptoms)	unknown	300 ppm-symptoms 200 ppm-increase in COHb level[b]	unknown	unknown
Ingestion	unknown	unknown	unknown	unknown	unknown	unknown
Carcinogenesis	unknown	unknown	unknown	unknown	unknown	unknown
Mutagenesis	S. thyphimurium TA 100	S. thyphimurium TA 100	positive[c] unknown	negative in Drosophila	unknown	unknown
Teratogenesis or Feto-toxicity:	unknown	unknown	unknown	feto-toxic	unknown	unknown
Environmental Presence						
Atmosphere (from Table 6.13)	0.75 ppb (typical)	unknown	0.02 ppb (typical)	unknown	unknown	unknown
Drinking Water (from Table 6.6)	unknown	unknown	unknown	1 µg/l (avg) 7 µg/l (max)	unknown	unknown
Food Supplies (from Table 6.18)	unknown	traces	unknown	unknown	unknown	unknown

Characteristics	Chloroform (CHCl$_3$)	Bromoform (CHBr$_3$)	Idoform (CHI$_3$)	Carbon Tetrachloride (CCl$_4$)	Carbon Tetrabromide (CBr$_4$)	Dibromo-chloro-methane (Br$_2$CHCl)	Bromo-dichloro-methane (BrCHCl$_2$)
TLV-TWA[a]	10 ppm (proposed)	0.5 ppm (skin)	0.2 ppm	10 ppm (skin)	0.1 ppm	none	none
Acute Toxicity	moderately toxic	highly toxic	unknown	highly toxic	unknown	unknown	unknown
Minimum Acute Toxic Doses							
Inhalation	~1,000 ppm (mild symptoms)	unknown	unknown	20-100 ppm (mild symptoms)	unknown	unknown	unknown
Ingestion	30-100ml (severe symptoms)	unknown	unknown	5-10 ml (severe, fatal)	unknown	unknown	unknown
Carcinogenesis	positive[d]	unknown	unknown	positive[e]	unknown	unknown	unknown
Mutagenesis	unknown	Ames Salmonella S. thyphimurium	unknown	possibly	unknown	unknown	unknown
Teratogenesis or Feto-toxicity:	feto-toxic	unknown	unknown	negative	unknown	unknown	unknown
Environmental Presence							
Atmosphere (from Table 6.13)	0.20 ppb (typical)	unknown	unknown	0.15 ppb (typical)	unknown	unknown	unknown
Drinking Water (from Table 6.6)	<0.1-366 µg/l	<0.8-92 µg/l	unknown	<2-3 µg/l	unknown	0.4-110 µg/l	0.3-116 µg/l
Food Supplies (from Table 6.18)	1-150 µg/kg	unknown	unknown	0.1-18 µg/kg	unknown	unknown	unknown

[a] Threshold Limit Value-Time Weighted Average airborne concentration for 40-hr work week, ACGIH (1976). The value for methylene chloride comes from U.S. DHEW (1976).
[b] COHb = carboxyhemoglobin level.
[c] BD-strain rats and A/Heston mice (subcutaneous route).
[d] A-strain mice, B6C3F1 strain mice, and Osborne-Mendel rats.
[e] C3H, A, Y, C, L, and Buffalo strain mice; Japanese, Wistar, and Osborne-Mendel rats; Syrian golden hamsters.

SOURCE: See Appendix B.

established for various susceptibility levels. Should a true threshold exist--and results from radiation carcinogenesis and mutagenesis argue against such a likelihood--it is not likely that it could be established unequivocally with any animal experiment of reasonable size. In view of the lack of data to establish the existence of a threshold in man for the compounds in question, the prudent approach from a public health viewpoint is to use nonthreshold models for risk estimation. This should always be the case unless there is, in a given situation, proper scientific evidence to establish both the existence of a biological threshold and some method of determining it quantitatively for man.

The multistage model for carcinogenesis assumes that cancer is single cell in origin and that the cell has undergone a series of mutational steps. This model includes as a special case the one hit or linear model but is not restricted to it.

The probit-type models, which have been developed empirically, tend to lack an obvious biological basis. In general, they yield lower estimates of risk below the experimental dose range than are obtained under a one-hit model. The risk estimates that are described in the next section were based on a modification of the Armitage-Doll multistage carcinogenisis model discussed in Crump et al. (1976), Guess and Crump (1976), and Guess et al. (1977). Since this formulation includes linear and non linear components, it can reflect both linear and sigmoidal-shaped (like probit-type models) dose-response curves. Under this model the probability of response as a function of dose, $P(d)$, is related to the dose (total daily intake) by the equation

$$P(d) = 1 - \exp\{-(\lambda_0 + \lambda_1 d + \lambda_2 d^2 + \ldots + \lambda_k d^k)\}$$

where k represents the number of transitional events in the carcinogenic process for the particular carcinogen in question and the $\lambda_i (i=0,1,\ldots,k)$ are unknown negative parameters. Although there are many parameters in the multistage model, the parameter of primary concern is λ_1, or the linear coefficient. The reason for this is that $P(d)$ will be approximately linear $(\lambda_0 + \lambda_1 d)$ at low dose levels. It is shown theoretically in Crump et al. (1976) that linearity is a very good approximation to the dose response function at low dose levels. Thus, the feared conservatism of linearity is without real basis for those direct acting carcinogens that affect processes with nonzero background levels.

Low dose risk estimates are obtained in the following manner. First the animal bioassay data is fitted to the multistage model and estimated parameter values are obtained. If pharmacokinetic data is available adjustments are made in the model. The reason is that the model assumes that the dose rate is the level of the active product at the cell and not the administered level. Lacking pharmacokinetic data we simply assume that the two levels are directly proportional. Using the estimated model, the risk of carcinogenesis at low dose levels is available along with statistical upper bounds. Next it is necessary to extrapolate the low dose risk estimates for the experimental animal to man. This is the point where it is felt that the greatest errors are made (Hoel et al. 1975). Meselson (NRC 1975b) has looked at several compounds where dose information in man is available. Empirically he found that the more sensitive animal species were reasonable quantitative predictors of man.

Considerably more research is needed both on the empirical study of carcinogenesis dose response differences in various species and the pharmacokinetic study of species scale up to man for model compounds. Without these studies it is not really possible to say whether the extrapolation is likely to be correct within an order of magnitude.

Chloroform and Carbon Tetrachloride Risk Estimation

Recently, two studies of the carcinogenic effects of long term oral exposure to chloroform have been reported[1] (National Cancer Institute [NCI] 1976a). In the first study, chloroform was administered orally to both rats and mice. Dose-response relationships were found for epithelial tumors of the kidney and renal pelvis in the rats and hepatocellular carcinomas in the mice. In the second study, chloroform was administered at somewhat lower doses to several strains of mice, rats, and dogs. In general, the data from this study were negative, although there were a few indications of a dose response.

Each data set was used individually to estimate low-dose effects. The method was to extrapolate the dose response curve to the low dose region by fitting a multistage model to the data (Guess and Crump 1976). A statistical upper 95 percent confidence interval was estimated. Next, the estimated low-dose risk was converted to human lifetime risk by using a per surface area adjustment (Hoel et al. 1975).

Using an oral dose of 1 µg/day, the lifetime human risks shown in Table 6.22 were calculated for this study. From these calculations, it is probably reasonable to use 2×10^{-7} or 0.2 in a million, as the estimated lifetime risk and

TABLE 6.22 Low-Dose[a] Lifetime Risk Estimates for 70 kg Man from Chloroform Based on Rodent Studies

	Risk Estimate	Upper 95% Confidence Estimate of Risk
NCI Study		
Rat male (kidney)	1.5×10^{-7}	3.1×10^{-7}
Rat female (kidney)	1.7×10^{-8}	8.5×10^{-8}
Mouse male (liver)	poor fit	1.3×10^{-6}
Mouse female (liver)	1.7×10^{-6}	2.2×10^{-6}
Roe Study		
Rat female (total)	3×10^{-7}	7.5×10^{-7}
Mouse male CF/1 (liver)	5.5×10^{-8}	4.9×10^{-7}
Mouse male swiss (liver)	2.3×10^{-7}	7.1×10^{-7}

[a] 1 μg/day oral dose.

SOURCE: Panel estimates based on National Cancer Institute (1976a) and Roe (1976)[1].

5(2×10^{-7}), or 1.0 in a million, as the upper 95 percent confidence bound.

Tardiff (1977) also estimated human cancer risk from exposure to chlorinated drinking water using only the NCI data. He calculated that a daily dose of 0.01 mg/kg would correspond to a risk estimate of 0.16 to 8.40 cancers per ten million population per year, depending on the animal species and statistical model used for extrapolation. When the above risk estimates for the NCI rat kidney data are compared to the linear model estimates derived by Tardiff (which is the only extrapolation that is really common to both studies), the results agree within a factor of two. This slight discrepancy is undoubtedly a reflection of the fact that calculations cited above incorporated a species conversion factor, while Tardiff's did not.

Since the dose response curve derived from the model at these low-dose levels is essentially linear (Crump et al. 1976), twice the dose of 1 µg/day would yield twice the risk and so on. Application of the risk estimates to total fluid intake values at estimated environmental exposure levels supplies data on lifetime risks for 70 kg man from chloroform in fluids, as shown in Table 6.23. These lifetime risks correspond to three different fluid intakes: 1.0, 3.7, and 1.95 liters/day as minimum, maximum and reference man values. These are actually the total fluid intakes shown in Tables 6.1 and 6.2. In calculating the risks it is conservative to assume that all the fluids contain chloroform at the indicated concentrations shown in Table 6.23. The same approach is used in calculating the lifetime risks for carbon tetrachloride exposure from fluids in Table 6.24. The calculated upper 95 percent bound on the risk calculation would increase these estimates by a factor of 5.

If one knew how to accurately subdivide the U.S. population of about 200 million people into appropriate exposure subgroups, it would be possible to obtain a rough approximation of the annual number of cancer deaths likely to result from exposure to chloroform in fluid intake from these lifetime risk estimates. For example, if 10 million males with total fluid uptakes like that of reference man were exposed to daily chloroform (fluid) levels of 21 µg/l (median environmental exposure concentration, Table 6.7), and if the average lifespan of this cohort were 70 years, then the expected annual number of cancer deaths in the cohort would be:

(estimated lifetime risk)(population at risk) x (1/average lifetime), or (8.1×10^{-6})(10^7)(1/70) = 1.16 cancer deaths/yr.

TABLE 6.23 Lifetime Risk for 70 kg Man from Chloroform in Fluids (2×10^{-7} for 1 µg/day)

	Min. Fluid Intake (1.0 l/day)	Max. Fluid Intake (3.7 l/day)	Reference Man Intake (1.95 l/day)
Minimum Exposure (0.1 µg/l)	2×10^{-8}	7.4×10^{-8}	3.9×10^{-8}
Median Exposure (21 µg/l)	4.2×10^{-6}	1.6×10^{-5}	8.1×10^{-6}
Maximum Exposure (366 µg/l)	7.3×10^{-5}	2.7×10^{-4}	1.4×10^{-4}

TABLE 6.24 Lifetime Risk for 70 kg Man from Carbon Tetrachloride in Fluids (5×10^{-8} for 1 µg/day)

	Min. Fluid Intake (1.0 l/day)	Max. Fluid Intake (3.7 l/day)	Reference Man Intake (1.95 l/day)
Minimum Exposure (2 µg/l)	1×10^{-7}	3.7×10^{-7}	1.9×10^{-7}
Maximum Exposure (3 µg/l)	1.5×10^{-7}	5.5×10^{-7}	2.9×10^{-7}

Using the inhalation tables presented early in this chapter (Table 6.14), a lifetime risk calculation based on oral doses can also be made for inhaled chloroform, as shown in Table 6.25. However, it must be stressed that chloroform has not been studied for carcinogenesis by inhalation. With a different route of exposure, there could be effects on the lung and other organs before chloroform arrives at the liver.

Carbon tetrachloride has been given orally in a number of studies in mouse, rat, hamster, and dog (International Agency for Research on Cancer 1972). In these studies, either the length of the study was too short or the dose level was too high for dose-response estimation of lifetime exposure. Recently carbon tetrachloride was used as a positive control in a lifetime carcinogenesis study of trichloroethylene (National Cancer Institute 1976b). Using the hepatocellular carcinoma data from rats in this study, the risk estimates for a daily intake of 1 µg by a 70 kg man, based on both male and female rat data, are approximately 5×10^{-8} (or 0.05 in a million) over an average lifetime and 1.5×10^{-7} (or 0.15 in a million) for the upper 95 percent confidence estimate. The uptake values can be converted to lifetime carcinogenesis risks as was done with chloroform, shown in Tables 6.24 and 6.26. For the upper 95 percent bounds, the risks are multiplied by 3.

Examination of the values shown in Tables 6.23-26 leads to these conclusions:

• For carbon tetrachloride and chloroform, the risk of cancer from inhalation and oral exposure should be of equal concern. (The maximum inhalation exposures for CCl_4 which are due to unusually high levels of CCl_4 measured in Bayonne, New Jersey indicate a greater concern in that case than with oral exposure.) This should stimulate animal experimentation using the inhalation route.

• The cancer risk from environmental exposures to chloroform is generally a greater health concern than that from exposure to carbon tetrachloride, but both appear to be small.

Epidemiological Investigations of Halomethanes in Drinking Water

Public concern about the health effects of contaminants in drinking water supplies was aroused in November 1974 by the release of an Environmental Defense Fund (EDF) report (Harris 1974), which was subsequently published in the scientific literature (Page et al. 1976). While the authors

TABLE 6.25 Lifetime Risk for 70 kg Man from Inhaled Chloroform[a]

	Min. Exposure	Typical Exposure	Max. Exposure
Min. Absorption	2.2×10^{-7}	2.2×10^{-6}	1.7×10^{-4}
Max. Absorption	3.5×10^{-7}	3.5×10^{-6}	2.6×10^{-4}

[a]Calculated by multiplying estimates of chloroform uptake from the outdoor atmosphere (Table 6.14) by 2×10^{-7}, the lifetime risk estimate for a daily oral dose of 1 µg chloroform.

TABLE 6.26 Lifetime Risk for 70 kg Man from Inhaled Carbon Tetrachloride[a]

	Min. Exposure	Typical Exposure	Max. Exposure
Min. Absorption	4.9×10^{-7}	6.1×10^{-7}	7.4×10^{-5}
Max. Absorption	5.6×10^{-7}	7.1×10^{-7}	8.5×10^{-5}

[a]Calculated by multiplying estimates of carbon tetrachloride uptake from the outdoor atmosphere (Table 6.15) by 5×10^{-8}, the lifetime risk estimate for a daily dose of 1 µg carbon tetrachloride.

did not specifically investigate the halogenated methane levels in water supplies, they did determine that there seems to be an association between total, urinary, and gastrointestinal cancer mortality rates in various Louisiana parishes and the drinking water obtained from the Mississippi River. De Rouen and Diem (1975) reviewed the EDF report and concluded that the evidence which it presented was not as strong as the publicity surrounding its release had suggested. They also hypothesized that ethnic and geographic differences might account for the apparent water source associations observed in the EDF report.

In a document prepared for the Cincinnati Board of Health, Buncher (1975) noted that a preliminary correlational analysis of county-wide, site-specific cancer mortality rates and drinking water concentrations of six volatile organic chemicals established the following associations:

• There was a significant positive correlation between chloroform levels and white male pancreatic cancer mortality;

• There was a significant positive correlation between chloroform levels and both white male and female bladder cancer mortality in selected urban counties; and

• The concentrations of the trihalomethanes were intercorrelated.

Three separate detailed epidemiological studies, based on the same data sets (or sets similar to those) examined by Buncher, have been conducted by D'Arge and Thanavibulchai (mimeograph 1976, Preliminary investigation of the relationship between cancer mortality and halogenated methanes in finished drinking water, University of Wyoming, Laramie), Cantor et al. (on detail to the National Cancer Institute from the U.S. Environmental Protection Agency, unpublished, 1977) and Hogan et al. (National Institute of Environmental Health Sciences, unpublished, 1977). While some positive associations were noted, results from various analyses were not always consistent across studies. (As is indicated by Hogan et al., the observed inconsistencies may be at least in part a reflection of the fact that an ecological methodology is being employed.) Furthermore, the main target organs in the animal studies (i.e., the liver and kidney) did not appear to be the most severely affected organs in humans.

Because of the lack of individual exposure data, the relative lack of stability of water data, the inability to account properly for population mobility patterns, the use of current rather than the more appropriate historical exposure data, and the almost certain lack of specificity of the statistical models employed, analyses based on indirect or ecological data are useful primarily for hypothesis generation and not for quantitative risk assessment. To that end the above cited studies tend to be supportive of whatever concern is generated by animal experimentation since they display an apparent pattern of association between contamination levels of drinking water and cancer mortality, albeit a somewhat obtuse pattern. It is unlikely that these types of epidemiological studies are sufficiently sensitive to detect human risk at the levels estimated from the animal studies, even should such a risk really exist.

Finally, two preliminary studies of chlorinated water supplies based on individual death certificate data have recently been completed. The first of these studies by Alavanja et al. at Columbia University School of Public Health (U.S. EPA 1976) compared all female gastrointestinal and urinary tract cancer deaths in seven New York counties for the years 1968 to 1970 with a corresponding set of matched controls. Alvanja et al. performed a chi-squared analysis of their data which indicated that females potentially exposed to chlorinated water were at higher risk than females residing in areas not supplied with chlorinated water, even when the comparison was restricted to urban areas only.

The second study by Kruse at John Hopkins (U.S. EPA 1977) considered all white, Washington County, Maryland liver and kidney cancer deaths during the interval from 1963 to 1974. A binary multiple regression technique attributed to Feldstein (1966) was used to correct the possibly confounding effects of such factors as sex, age, marital status, smoking history, residence history, and so on. The author concluded that "...chloroform status of water consumed was observed to have no effect on either outcome or interest [liver and kidney cancer mortality], although the direction of the difference between the adjusted liver cancer incidence rates is consistent with the hypothesis that chlorinated water drinkers are at a greater risk than unchlorinated water drinkers." When the analysis was restricted to white females, chlorinated water drinkers were at (a statistically non-significant) excess risk for both types of cancer. Kruse regards these findings as sufficient to justify the initiation of a case-control study.

While both of these studies have technical shortcomings, they do represent a more direct attempt to assess the relative human cancer risk associated with exposure to chlorinated drinking water. The Alavanja and Kruse studies are suggestive of a possible relationship between chlorinated water, chloroform in water, and human cancer. To clarify this additional studies are required.

NOTE

1   Roe, F.J.C. (1976) Preliminary report of long-term tests of chloroform in rats, mice, and dogs. (Unpublished results)

REFERENCES

American Conference of Governmental Industrial Hygienists
(ACGIH) (1976) TLVs^R Threshold Limit Values for Chemical
Substances and Physical Agents in the Workroom
Environment with Intended Changes for 1976. Cincinnati,
Ohio: American Conference of Governmental Industrial
Hygienists.

Bauchop, T. (1967) Inhibition of rumen methanogenesis by
methane analogues. Journal of Bacteriology 94:171-175.

Bellar, T.A., J.J. Lichtenberg, and R.C. Kroner (1974)
Occurrence of organohalides in chlorinated drinking
waters. Journal of the American Water Works Association
66 (12):703-706.

Buncher, C.R. (1975) Cincinnati Drinking Water--An
Epidemiologic Study of Cancer Rates. Report for the
Cincinnati Board of Health. November. Cincinnati, Ohio:
Cincinnati Board of Health. (Shortened version, American
Journal of Public Health 67:725-729, 1977.)

Chopra, N.M. and L.R. Sherman (1972) Systemic studies on the
breakdown of p,p¹-DDT in tobacco smokes. Analytical
Chemistry 44:1036-1038.

Corn, M. (1974) Characteristics of tobacco sidestream smoke
and factors influencing its concentration and
distribution in occupied spaces. Scandinavian Journal of
Respiratory Diseases (Suppl.) 91:21-36.

Crump, K.S., D.G. Hoel, C.H. Langley, and R. Peto (1976)
Fundamental carcinogenic processes and their
implications for low dose risk assessment. Cancer
Research 36 Part 1, (9):2973-2979.

De Rouen, T.A. and J.E. Diem (1975) The New Orleans drinking
water controversy, a statistical perspective. American
Journal of Public Health 65(10):1060-1062.

Dowty, B., D. Carlisle, J.L. Laseter, and J. Storer (1974)
Halogenated hydrocarbons in New Orleans drinking water
and blood plasma. Science 187:75-77.

Feldstein, M.S. (1966) A binary variable multiple regression method of analysing factors affecting peri-natal mortality and other outcomes of pregnancy. J. Royal Statistical Society 129 (Series A, Part 1):61-73.

Guess, H.A. and K.S. Crump (1976) Low dose extrapolation of data from animal carcinogenicity experiments--analysis of a new statistical technique. Mathematical Biosciences 32(1-2):15-36.

Guess, H., K. Crump, and R. Peto (1977) Uncertainty estimates for low dose rate extrapolations of animal carcinogenicity data. Cancer Research 37(10):3475-3483.

Harris, R.H. (1974) The Implications of Cancer-Causing Substances in Mississippi River Water. Washington, D.C.: The Environmental Defense Fund.

Hoel, D.G., D.W. Gaylor, R.L. Kirschstein, U. Saffiotti, and M.W. Schneiderman (1975) Estimation of risks of irreversible, delayed toxicity. Journal of Toxicology and Environmental Health 1:133-151.

International Agency for Research on Cancer (1972) IARC Monograph on the Evaluation of Carcinogenic Risk of Chemicals to Man, Vol. 1. Lyon: World Health Organization.

International Commission for Radiological Protection (ICRP) (1975) Report of the Task Group on Reference Man, edition no. 23. New York: Pergamon Press.

Jackson, S. and V.M. Brown (1970) Effect of toxic wastes on treatment processes in watercourses. Water Pollution Control 69:292-302.

Jones, J.R.E. (1947a) The oxygen consumption of Gasterosteus acculeatus L. in toxic solutions. Journal of Experimental Biology 23:298.

Jones, J.R.E. (1947b) The reactions of Pygosteus pungitius L. in toxic substances. Journal of Experimental Biology 24:100.

Lehmann, K.B. and Hasegawa (1910) Studies of the absorpotion of chlorinated hydrocarbons in animals and humans. Archiv fur Hygiene 72:327-342. (U.S. Department of Health, Education, and Welfare Publication No. [NIOSH] 75-114.)

Lehmann, K.B. and L. Schmidt-Kehl (1936) The thirteen most important chlorinated aliphatic hydrocarbons from the standpoint of industrial hygiene. Archiv fur Hygiene 116:131-268 (U.S. Department of Health, Education, and Welfare Publication No. [NIOSH] 76-138 and 76-133.)

Lillian, D., H.B. Singh, A. Appleby, L. Lobban, R. Arnts, R. Gumpert, R. Hague, J. Toomey, J. Kazazis, M. Antell, D. Hansen, and B. Scott (1975) Atmospheric fates of halogenated compounds. Environmental Science and Technology 9(12):1042-1048.

McConnell, G., D.M. Ferguson, and C.R. Pearson (1975) Chlorinated hydrocarbons and the environment. Endeavor 34(121):13-18.

McKee, J.E. and H.W. Wolf (1971) Water Quality Criteria. Sacramento, Calif.: The Resources Agency of California, State Water Resources Control Board, Publication No. 3-A: USPHS, HEW.

National Cancer Institute (1976a) Report on the Carcinogenesis Bioassay of Chloroform. Washington, D.C.: U.S. Government Printing Office; Springfield, Va.: National Technical Information Service.

National Cancer Institute (1976b) Carcinogenesis Bioassay of Trichloroethylene. NCI-CG-TRA-2. Washington, D.C.: U.S. Government Printing Office; Springfield, Va.: National Technical Information Service.

National Research Council (1975a) Environmental Impact of Stratospheric Flight: Biological and Climatic Effects of Aircraft Emissions in the Stratosphere. Climatic Impact Committee. Washington, D.C.: National Academy of Sciences.

National Research Council (1975b) Contemporary Practices and Prospects: The Report of the Executive Committee. Volume I. Pest Control: An Assessment of Present Alternative Technologies. Environmental Studies Board. Washington, D.C.: National Academy of Sciences.

National Research Council (1976b) Halocarbons: Environmental Effects of Chlorofluoromethane Release. Committee on Impacts of Stratospheric Change. Washington, D.C.: National Academy of Sciences.

National Research Council (1976a) Halocarbons: Effects on Stratospheric Ozone. Panel on Atmospheric Chemistry. Washington, D.C.: National Academy of Sciences.

Owens, D.F. and A.T. Rossano (1969) Design procedures to control cigarette smoke and other air pollutants. ASHRAE Trans. 75:93-102.

Page, T., R.H. Harris, S.S. Epstein (1976) Drinking water and cancer mortality in Louisiana. Science 193:55-57.

Pearson, C.R. and G. McConnell (1975) Chlorinated C_1 and C_2 hydrocarbons in the marine environment. Proceedings of the Royal Society of London, Series B 189(1096):305-322.

Philippe, R.J. and M.E. Hobbs (1956) Some components of the gas-phase of cigarette smoke. Analytical Chemistry 28:2002-2006.

Ramanathan, V. (1975) Greenhouse effect due to chlorofluorocarbons: Climatic implications. Science 190(4209):50-52.

Sanwick, J.D. and M. Foulkes (1971) Inhibition of anaerobic digestion of sewage sludge by chlorinated hydrocarbons. Water Pollution Control 70:58-70.

Stickley, D.P. (1970) The effect of chloroform in sewage in the production of gas from labor digestors. Water Pollution Control 69:585-592.

Symons, J.M. (1976) Interim Treatment Guide for the Control of Chloroform and Other Trihalomethanes. Cincinnati, Ohio: U.S. Environmental Protection Agency.

Symons, J., T.A. Bellar, J.K. Carswell, J. DeMarco, K.L. Kropp, G.G. Robeck, D.R. Seeger, C.J. Slocum, B.L. Smith, and A.A. Stevens (1975) National organics reconnaissance survey for halogenated organics (NORS). Journal of the American Water Works Association 67:634-647.

Tardiff, R.G. (1977) Health effects of organics: Risk and hazard assessment of ingested chloroform. Journal of the American Water Works Association 69:658-662.

Thiel, P.G. (1969) The effect of methane analogues on methanogenesis in anaerobic digestion. Water Research 3:215-223.

U.S. Department of Health, Education, and Welfare (1974) Criteria for a Recommended Standard. Occupational Exposure to Chloroform. HEW Publication No. (NIOSH) 75-114. Washington, D.C.: U.S. Government Printing Office.

U.S. Department of Health, Education, and Welfare (1975) NIOSH Criteria for a Recommended Standard. Occupational Exposure to Carbon Tetrachloride. HEW Publication No. (NIOSH) 76-133. Washington, D.C.: U.S. Government Printing Office.

U.S. Department of Health, Education, and Welfare (1976) NIOSH Criteria for a Recommended Standard. Occupational Exposure to Methylene Chloride. HEW Publication No. (NIOSH) 76-138. Washington, D.C.: U.S. Government Printing Office.

U.S. Department of Transportation (1975) The Impacts of Climatic Changes in the Biosphere. Climatic Impact Assessment Program (CIAP) Monograph 5, Part I: Ultraviolet Radiation Effects, NTIS No. PB 247/724; Part 2: Climatic Effects, NTIS No. PB 247/725. Springfield, Va.: National Technical Information Service.

U.S. Environmental Protection Agency (1975) Preliminary Assessment of Suspected Carcinogens in Drinking Water, and Appendices. A Report to Congress. December 1975. Washington, D.C.: U.S. Environmental Protection Agency.

U.S. Environmental Protection Agency (1976) Report of Case Control Study of Cancer Deaths in Four Selected New York Counties in Relation to Drinking Water Chlorination. EPA Contract No. 76224. Cincinnati, Ohio: Health Effects Research Lab, U.S. Environmental Protection Agency.

U.S. Environmental Protection Agency (1977) Chlorination of Public Water Supplies and Cancer: Washington County, Maryland Experience. Preliminary Report. Cincinnati, Ohio: Health Effects Research Lab, U.S. Environmental Protection Agency.

Wang, W.C., Y.L. Yung, A.A. Lacis, T. Mo, and J.E. Hansen (1976) Greenhouse effects due to man-made perturbations of trace gases. Science 194(4266):685-690.

Wood, J.M., F.S. Kennedy, and R.S. Wolfe (1968) The reaction of multihalogenated hydrocarbons with free and bound reduced Vitamin B-12. Biochemistry 7:1707-1713.

CHAPTER 7

CONTROL TECHNIQUES, OPTIONS, AND COSTS

In considering possible control techniques it is most fruitful to relate the potentially adverse effect to the route of exposure and then to the source of the compound responsible for that exposure. The next step is to identify control techniques that will be effective in reducing exposure and finally the costs and benefits associated with the control techniques are determined.

CARBON TETRACHLORIDE

The adverse effects of carbon tetrachloride (CCl_4) include its potential carcinogenicity to humans and the possible alteration of stratospheric ozone levels by the continued buildup of the compound in the lower atmosphere.

Human Exposure

Table 6.20 summarizes human exposure to CCl_4 and shows that, at typical environmental levels, inhalation of outdoor air is slightly more important than ingestion of food and fluids in the total human exposure to this compound. In certain indoor situations and at isolated outdoor locations, there may be somewhat higher human exposure to CCl_4 from inhalation than is the case in the typical outdoor situation.

Atmospheric levels of CCl_4 seem to be related to the total environmental input of the compound from all sources; control techniques that reduce the entry of CCl_4 into any compartment of the environment will reduce its atmospheric levels. But even when controls are employed, it will take a number of years to produce a significant drop in human exposure through inhalation because of the long residence time of this compound in the atmosphere.

The levels of CCl_4 in the human diet are much higher than would be predicted from the equilibrium distribution of CCl_4 between air and pure water, and the source of CCl_4 in water and food is still obscure. The fumigation of grain with CCl_4 makes a small contribution to the dietary intake of CCl_4 by humans, although the impact of fumigation on the total environment, especially the atmospheric levels of CCl_4 especially, may be more significant. Water contamination could result from the discharge of CCl_4 by industrial plants or in miscellaneous uses of the compound. A linkage between these sources and the levels of CCl_4 in drinking water has not been established, however, and this limits the range of control options that are available for reducing human exposure to CCl_4 through food and fluids.

Stratospheric Exposure

The transfer of CCl_4 to the stratosphere is in direct proportion to the tropospheric concentration of the compound which, in turn, is determined by the total amount of CCl_4 released to or formed in the environment. Control techniques to reduce the exposure of the stratosphere to CCl_4 thus must relate to the total environmental entry of the compound.

Control Techniques

The following are possible methods of controlling the entry of CCl_4 to the environment:

- Reduce emissions to air and wastewater streams from the industrial manufacture and use of CCl_4.

- Ban the use of CCl_4 as a grain fumigant.

- Ban the production of fluorocarbons F-11 and F-12, which are produced by the hydrofluorination of CCl_4.

- Eliminate other uses of CCl_4. Too little information is available on the nature and importance of these uses to make possible an evaluation of this control option at this time.

- Prohibit discharge of perchloroethylene if further research shows that atmospheric degradation of this compound is producing significant amounts of CCl_4. This control option would involve a major disruption in the U.S. economy because perchloroethylene is used extensively as a solvent in dry cleaning.

The reduction of industrial emissions to air and water, the first of these control methods, can be brought about in a number of ways, with varying degrees of effectiveness and cost.

Refrigeration effectively reduces CCl_4 emissions in streams of concentrated gas; however, very high costs and energy consumption are involved in this technique. High temperature or catalytic incineration can also be used to reduce emissions to the atmosphere, as these methods will destroy CCl_4 in gas streams (Burgess 1976). These methods may be useful when the concentration of unwanted halocarbons is fairly high (>100 ppm). Although the cost of this approach is prohibitive for small installations, the technique may be useful for large plants. Costly corrosion-resistant construction materials and effective scrubbers are required to handle combustion products such as Cl_2 and HCl in wet systems.

Carbon adsorption can be used to remove CCl_4 from concentrated air streams, but the loading capacity is low, making it cost effective only for relatively high concentrations of CCl_4 (>200 ppm). Polymeric sorbents are more cost effective than activated carbon in air streams containing high concentrations. Once adsorbed the compound will desorb readily into regenerant streams. An important consideration when adsorption methods are used is the proper disposal of the exhausted adsorbent that has been saturated with CCl_4. Regeneration may not be practical, and if the adsorbant is incinerated this must be done under carefully controlled conditions. Direct water scrubbing of airstreams is impractical because of the limited solubility of CCl_4.

Losses from plants where CCl_4 is produced or used as a chemical intermediate usually occur in dilute gas streams (<10 ppm) and are largely of the fugitive type. These losses take place when reactors are opened for maintenance, when there are leaks in pump seals or when there are other plant malfunctions, during chemical transfers, and at various other times. It may be possible to reduce CCl_4 emissions through better design and control of the manufacturing processes, but the results will greatly depend upon the individual situation. Further investigation of this approach to reducing the entry of CCl_4 to the atmosphere requires a case-by-case study of each emission source.

Air- or steam-stripping is the simplest technique for removing CCl_4 from contaminated water (TRW Systems Group 1975). Final CCl_4 concentrations of about 1 mg/l in the effluents can be achieved at reasonable cost, i.e., the product recovered pays for the investment in and operation of the control equipment. If steam is used, most of the CCl_4 can be recovered from the air stream.

Carbon adsorption or sorption using appropriate organic polymeric materials is another method for removing halomethanes from water. Carbon adsorption is partially effective; however, loadings of CCl_4 are low and the associated costs are usually high, even with reimbursable credits for recovered materials. Incineration of aqueous streams is impractical because the energy demands are great.

Costs associated with each of the techniques for reducing CCl_4 emissions in air and wastewater streams are highly dependent on individual circumstances. A major effort by industrial producers and users of CCl_4 would therefore be required to define the costs that are involved in gaining significant reductions in CCl_4 emissions to the environment from their operations.

Patterns of CCl_4 use and emission could be altered drastically by curtailing certain direct dispersive uses of the compound, such as fumigation, and by substituting fluorocarbons that use chloroform as a raw material for those that use CCl_4.

Currently, the major dispersive use of CCl_4 is in the fumigation of grain for export. Although this use comprises less than 5 percent of the annual CCl_4 production, it accounts for about half of all CCl_4 emissions to the environment. Alternative compounds that might be used for fumigating grain include carbon bisulfide, ethylene dichloride, ethylene dibromide, methyl bromide, phosphine, and chloropicrin. The disadvantages of these fumigants are higher costs, lower efficacy, greater precautions required in handling the materials, a slower release of residuals, and ineffectiveness at lower temperatures. It is beyond the scope of this Panel to analyze the feasibility of replacing CCl_4 with other fumigants or to assess the relative toxicities and the costs involved in the use of alternative compounds, but such an analysis must be made in order to evaluate this course of action.

A significant reduction in CCl_4 emissions could result from substitution of other materials for fluorocarbons F-11 and F-12 (trichlorofluoromethane and dichlorodifluoromethane), which are produced by hydrofluorination of CCl_4. A considerable amount of research would be required to find substitutes that are effective, safe, and economically feasible. However, if such substitutes were available, most of the emissions of carbon tetrachloride from its production and use as an intermediate (about 30 percent of the total emission) could be eliminated.

Techniques for removal of low levels of CCl_4 in fluids and food have not been studied in detail. The high vapor pressure and low water solubility of CCl_4 suggest that the compound could be removed by transfer to the atmosphere. Available data on loss of CCl_4 from foods treated with the compound confirm that volatilization does occur (see Chapter 5). The next step would be to determine the time, temperature, air movement, and other factors that are necessary for effective removal as well as the costs and other side effects that would result from removal efforts. Removal of CCl_4 from water and water-based fluids could be partially accomplished by carbon adsorption.

CHLOROFORM AND OTHER TRIHALOMETHANES

The adverse effects upon humans and the environment that have been identified for chloroform ($CHCl_3$) are similar to those for carbon tetrachloride: the potential carcinogenicity of the compound to humans, and the possible alteration of stratospheric ozone levels. The atmospheric background level of $CHCl_3$, 20 to 40 parts per trillion, and the compound's relatively short tropospheric lifetime, suggest, however, that the latter concern should not receive high priority; control techniques relating to the tropospheric burden of $CHCl_3$ are therefore not considered here.

Data have not been generated that show the potential adverse effects on human health from dibromochloromethane, bromodichloromethane, and bromoform. These compounds are formed along with chloroform during the chlorination of potable water, which is probably the only significant way that humans are exposed to them. The control of $CHCl_3$ in drinking water would also result in control of these compounds.

Human Exposure

Major routes of human exposure to $CHCl_3$ are through fluid intake and inhalation, as shown in Table 6.20. Control techniques relating to the disinfection of water supplies by chlorination are likely to have significant effects on human exposure to $CHCl_3$ through fluid intake because the presence of $CHCl_3$ in water, and possibly in other fluids and foods, is primarily due to this disinfection practice. As noted earlier, there is preliminary evidence that exposure to $CHCl_3$ through inhalation may be significantly affected by the combustion of leaded gasoline in automobiles, which may contribute to the elevated ambient concentrations of $CHCl_3$ in urban areas.

Mass balance analysis suggests that a combination of nonanthropogenic sources and secondary formation from anthropogenic sources must account for the major burden of $CHCl_3$ in the troposphere (see Table 5.23).

Control Techniques

Control techniques that could be considered for chloroform include the following:

- Control the amount of $CHCl_3$ in drinking water.

- Reduce emissions of $CHCl_3$ in its industrial production and use.

- Prohibit discharge of trichloroethylene, if further research shows that $CHCl_3$ is produced in significant quantities from atmospheric degradation of that compound.

- Remove chlorinated compounds from gasoline to prevent the introduction of $CHCl_3$ into the atmosphere from automotive exhausts.

The last three of these control techniques relate to background and urban tropospheric concentrations of $CHCl_3$. Before control options relating to the presence of $CHCl_3$ in the atmosphere are examined in more detail, however, it will be necessary to develop a better understanding of the relationship between the sources of $CHCl_3$ and its atmospheric levels. If further study shows, for example, that the high urban levels of $CHCl_3$ are related to the use of chlorinated compounds in leaded gasoline, an appropriate control option probably would be to remove such compounds from gasoline, and the economic implications of this action should be investigated at this time. The chlorination of drinking water and industrial wastewater also may result in $CHCl_3$ losses to the atmosphere, but more information is needed before the atmospheric implications of controls placed on chlorination can be usefully discussed.

A direct relationship has been established between human exposure to $CHCl_3$ and the use of chlorine for disinfection. Control techniques and associated costs for reducing human exposure via this route are therefore examined in some detail.

Control of Chloroform in Drinking Water

Certain trihalomethanes, i.e., bromoform, bromodichloromethane, dibromochloromethane, and chloroform have been found in finished drinking waters (Rook 1976, Stevens et al. 1975). Detection resulted from the development of sophisticated analytical techniques and not from any major changes in water treatment practices. Chlorine has been shown to react with organic precursors that are present in raw water, particularly surface waters,

to form trihalomethanes; it is possible that there are other
by-products as well (Morris and McKay 1975). The formation
of trihalomethanes is influenced by the concentrations of
precursors, the form of chlorine used as a disinfectant, pH,
and temperature. Preliminary investigations of alternative
water treatment processes include techniques such as: (a)
reducing or eliminating certain precursors of the
trihalomethanes before chlorination; (b) changing the
disinfectant or its point of application; and (c) removing
trihalomethanes after they have formed. Although specific
precursors have not been identified, humic and fulvic acids
and simple compounds containing acetyl groups are capable of
combining with chlorine to form trihalomethanes (Morris and
Johnson 1976).

Removal of Precursors

In attempts to reduce the organic precursors of
trihalomethanes before chlorination, ozone, permanganate,
and chlorine dioxide have been added to the raw water.
Ozone is selective in destroying humic acids and double-
bonded compounds, but unusually large concentrations of
ozone are required. Permanganate is a more general oxidant,
but reaction is slow. Moderate success has been obtained
using chlorine dioxide (Symons 1976), but the chlorine
produced as a by-product forms some trihalomethanes before
the precursors are destroyed. Combinations of these
oxidizing agents may be more effective than single agents
(Morris and McKay 1975).

Coagulation, sedimentation, and filtration processes
promote the removal of about 90 percent of the precursors to
trihalomethane formation (Symons 1976, Stevens et al. 1975).
The removal of precursors by these means is usually not
complete, and some formation of trihalomethanes (~20 µg/l)
may still occur during chlorination. The improved use of
chemical coagulants and flocculants may suffice to reduce to
tolerable levels the precursors that are in particulate form
or that are adsorbed onto particulates (Stevens et al.
1975).

Carbon adsorption is technically feasible but
economically uncertain as a technique for precursor removal.
Its advantages include the removal of portions of the
suspected precursors; removal of other organic materials
that may form unknown byproducts; removal of specific
organic materials that cause concern, such as pesticides and
those responsible for taste and odor problems; and removal
of portions of some trihalomethanes that may already be
present in raw water. Final trihalomethane concentrations
in the range of 5 to 10 µg/l (Love et al. 1975) could
probably be reached. The disadvantages of carbon adsorption

are the limitations in the adsorptive capacity of carbon for certain organics, the need for adequate pretreatment to prevent fouling of the carbon by suspended solids, the need for frequent regeneration of the carbon, the potential for introducing carbon particles or chemicals that have leached from the carbon and whose hazards are unknown into finished water, and significant operating and capital costs.

Powdered activated carbon has been reported to require "unrealistically" high concentrations for adsorption of trihalomethane precursors, whereas granular activated carbon was more effective (Symons 1976). This evaluation was limited, however, by the fact that two dissimilar carbons were compared on unequal bases. Further investigation of the capacities of other commercially available carbons is necessary before a decision on the effectiveness of the carbon adsorption method for removal of the precursors of trihalomethanes can be made.

Anion exchange following coagulation, sedimentation, and filtration of drinking water has also been evaulated (Rook 1976). This proved to be an expensive method for removing the organic precursors of trihalomethanes, since only about half the precursors were removed and frequent regeneration was required. Furthermore, the disposal of the desorbed organic materials during regeneration was not considered in the evaluation and no distinction was made between removal of precursors by resins or by chemical pretreatment. The anion exchange approach does not appear to be technically or economically feasible.

Change of Disinfectant or Point of Application

The second approach to reducing or eliminating the formation of trihalomethanes in drinking water is to change the disinfectant or its time of application. Possible substitutes for chlorine are ozone, chlorine dioxide, and chloramines (Sawyer 1976), but unfortunately, no other disinfectant offers all the advantages of chlorine. Ozone has been used effectively without forming trihalomethanes, but small doses of chlorine must be added to provide residual protection. Chlorine dioxide effectively disinfects drinking water and provides residual protection, but chlorine and other trihalomethanes are by-products. Chloramines do not produce trihalomethanes and do provide residual protection, but they are weak disinfectants compared to chlorine. The possibility that other unknown materials of concern may be formed through the use of alternative disinfectants exists and must be considered.

The time of chlorination may be important in reducing the ultimate concentration of trihalomethanes (Stevens et

al. 1975). Trihalomethane formation can be minimized if
chlorine is applied after coagulation, sedimentation, and
filtration of drinking water (Symons 1976). Sufficient
contact time for safe disinfection prior to distribution,
however, must be insured. Postchlorination could replace
prechlorination, but only when the source water is
relatively unpolluted and does not require intense
chlorination for long periods of time (Morris and McKay
1975). Problems with algae, slimes, and higher plant forms
may result if prechlorination does not take place.

Removal of Trihalomethanes

Ozonation, aeration, and carbon adsorption have been
tested for the removal of trihalomethanes after their
formation; these methods are generally more expensive than
prevention techniques. Impractically high concentrations of
ozone were required, for example, to achieve substantial but
incomplete removal of trihalomethanes in the ozonation
process.

Diffused-air aeration can be effective for the removal
of trihalomethanes with air-to-water ratios that are
considerably higher than those normally used to treat
drinking water. Air-to-water ratios of 30:1, 16:1, and 8:1
(v/v) can achieve chloroform removal efficiencies of about
90 percent, 80 percent, and 55 percent, respectively,
depending upon the initial concentration of chloroform
(Symons 1976). The intensity of aeration can be controlled
for maximum removal of chloroform, depending upon the
influent concentration. Disadvantages of diffused-air
aeration are the partial reduction in residual chlorine; the
increase in dissolved oxygen, which increases corrosion of
the distribution system; the formation of trihalomethanes
during distribution of drinking water as residual chlorine
and precursors react; and the removal, but not total
destruction, of trihalomethanes.

The best technique for removal of trihalomethanes after
their formation is activated carbon adsorption. Levels of
~ 0.1 $\mu g/l$ can be achieved by this method. Granular
activated carbon was found somewhat more effective than
powdered activated carbon in limited tests (Symons 1976).
Efficiencies of greater than 95 percent in trihalomethane
removal were achieved and maintained over a three-week
period (Rook 1976). Other uncharacterized organics are also
removed by carbon adsorption. The major disadvantage of
activated carbon adsorpton is the limited adsorptive
capacity which necessitates frequent reactivation of the
carbon. When adsorption is employed after chlorination,
prechlorination should be intense and the chlorine residual
must be substantial to encourage complete trihalomethane

formation, rather than the production of partially
chlorinated intermediates, which may be more dangerous and
more difficult to remove (Morris and McKay 1975).

The organic content of a source water, existing
treatment processes, and the size of a treatment facility
largely determine the applicability of various alternatives
for trihalomethane control. Chlorination of the highest
quality water after chemical treatment and improvement or
adaptation of existing facilities should be considered
first. The use of treatment techniques for the control of
concentrations of trihalomethane precursors or changing to
some disinfectant other than Cl_2 is less costly than removal
of trihalomethanes once they have been formed. No matter
what changes are considered, the microbiological quality of
drinking water must be carefully monitored and the benefits
should be carefully weighed before the employment of
alternatives with high capital and operating costs.

Costs of Removal

Because relatively limited information is available on
the costs of reducing or eliminating trihalomethanes in
drinking water, this discussion centers on three options:
(1) alteration of existing disinfection practices, (2)
removal of the precursors of trihalomethanes by granular
activated carbon adsorption, and (3) removal of
trihalomethanes by aeration. The cost of each option
depends in part upon establishing the degree of
trihalomethane removal that is required.

The implementation of trihalomethane control in
municipal water systems must apply to the entire water
supply, irrespective of whether the water is ultimately
consumed by humans. Increases in costs of treatment are
distributed to the public consumer in two ways: directly in
water bills, and indirectly in the price of goods purchased
from industries that use the more expensive water. Because
costs are borne in this manner, the total costs for the
volume of water that is treated must be prorated on a per
capita basis for the total volume of water that is treated,
even though only a fraction of the treated water is consumed
by individual households. The present average total
consumption of water in 49 U.S. cities that have water
treatment plants with a greater than 40 million gallons per
day potable water capacity (i.e., the 99 largest municipal
water treatment plants in the United States), is 200 gallons
per day per capita or 292,000 gallons per year for a family
of four (U.S. EPA 1976). This figure is used in the
calculations in this chapter, although values calculated for
per capita municipal water treatment volumes actually vary
from less than 100 gallons per day in some parts of the

United States to 300 gallons per day in other parts of the nation.

The cost of disinfection varies considerably with location, size of plant, and choice of disinfectant; generally the cost is about 10 percent of the total cost of water delivery. Estimated total base costs (operating and capital costs) of conventional disinfection by chlorination compared to the costs of disinfection with chlorine dioxide and by ozonation are shown in Table 7.1. These estimated costs assume that no disinfection facilities existed prior to installation of each, and they do not consider the cost of converting to an alternative disinfectant. For larger water treatment plants, the costs of chlorination and ozonation are roughly comparable, while these costs vary markedly in plants with a treatment capacity of a million gallons a day. The estimated incremental costs of disinfection that would be incurred by changing from a chlorine disinfection system to another system are shown in Table 7.2. An existing chlorination system could add a chlorine dioxide disinfection capability to small plants at a small incremental cost.

Table 7.3 shows the annual prorated costs of disinfection for an average household of four that consumes 200 gallons of water a day per capita. These figures are based on the treatment plant costs in Table 7.1 and they assume that households will ultimately bear the increased costs to industry as well as the costs of household water consumption.

Total costs of conventional water treatment, including operating costs and amortization of capital facilities for pumping, coagulation, flocculation, sedimentation, filtration, disinfection, and distribution, may amount to $50 to $500 per million gallons (American Water Works Association Industry Range Estimate, Denver, personal communication, 1977) or $15 to $150 per year for an average household. Control of trihalomethanes by changing disinfection processes could increase treatment costs by 15 percent.

Additional operating and capital costs would be incurred for the removal of trihalomethanes or their precursors by carbon adsorption or by aeration. The estimated additional costs of these methods are shown in Table 7.4. It has been implied in studies of the subject, but not directly stated, that a residual of trihalomethanes below the detectable limit of <0.1 µg/l of chloroform may be reached by control techniques. This amounts to a 99.5 percent removal from a median national water supply concentration of 20 µg/l (Symons 1976). Costs could be reduced if the residual requirement were set at 2 µg/l. Chemical

TABLE 7.1 Estimated Base Costs of Disinfection ($/$10^6$ gal)

Disinfection System	Design Capacity, 10^6 gal/day				
	1	5	10	100	150
Chlorination	$36	$16	$12	$ 7	$ 6
Chlorine dioxide	38	20	17	13	13
Ozonation (air)	63	23	16	8	8
Ozonation (oxygen)	77	25	18	9	8

SOURCE: Symons (1976).

TABLE 7.2 Estimated Incremental Costs of Disinfection ($/$10^6$ gal)[a]

Disinfection System	Design Capacity, 10^6 gal/day				
	1	5	10	100	150
Chlorine dioxide	$ 2	$4	$5	$6	$7
Ozonation (air)	27	7	4	2	2
Ozonation (oxygen)	41	9	6	2	2

[a] Assuming conversion is made from existing chlorine disinfection system.

SOURCE: Symons (1976).

TABLE 7.3 Prorated Costs of Disinfection ($/family/year)[a]

Disinfection System	Design Capacity, 10^6 gal/day	
	1	150
Chlorination	$10	$2
Chlorine dioxide	11	4
Ozonation (air)	18	2
Ozonation (oxygen)	22	2

[a] Assuming a family of four, 200 gal/day per capita consumption (calculated by multiplying values in Table 7.1 by a conversion factor of 0.292).

TABLE 7.4 Estimated Additional Costs of Trihalomethane Prevention or Removal ($/$10^6$ gal)

Control Option	Design Capacity, 10^6 gal/day				
	1	5	10	100	150
Prevention or Removal					
Granular activated carbon adsorption (contractors)	$463	$175	$136	$67	$63
Granular activated carbon adsorption (media replacement)	410	155	117	60	51
Removal					
Aeration					
(30:1 air-to-water ratio ~90 percent removal)	223	146	127	90	85
(16:1 air-to-water ratio ~80 percent removal)[a]	151	92	78	52	50
(8:1 air-to-water ratio ~55 percent removal)[a]	100	56	47	30	28

[a]Costs are derived from the cost of 90 percent removal of trihalomethanes on the basis of volume of air required for each air-to-water ratio and plant capacity (see section of Chapter 8 on "Costs of Removing Chloroform and Other Trihalomethanes from Drinking Water").

SOURCE: Symons (1976).

coagulation/flocculation and filtration before chlorination, followed by aeration, may be adequate at this level.

Two adsorption options are considered in Table 7.4: installation of carbon contactors in series with existing filters, and replacement of existing filter media with granular activated carbon. The costs of the two carbon adsorption options are similar. If the less costly of these options is selected--granular activated carbon adsorption with media replacement--total costs of water treatment will be increased by $51 per million gallons a day for large treatment plants to $410 per million gallons a day for small plants. This would be in addition to disinfection costs shown on Table 7.1, which range from a low of $6 per million gallons a day for chlorination in the largest treatment plants to the highest base cost of $77 per million gallons a day for ozonation, using oxygen, in the smallest plants. Control of trihalomethanes in potable water by the carbon adsorption method, therefore, may result in an ultimate cost borne by the average household of from $15 to $120 a year for the additional treatment. This would double the total cost of water treatment for the average household, as estimated on the basis of the cost data from the American Water Works Association.

Estimated costs for removing trihalomethanes by diffused-air aeration are also shown in Table 7.4. Three air-to-water ratios, which represent approximately 90 percent, 80 percent, and 55 percent removal of trihalomethanes, are considered. Based on costs given in Table 7.4, diffused-air aeration would represent additional annual costs to the average household, depending on treatment plant capacity, of $8 to $29 (for 55 percent removal of trihalomethanes), $15 to $44 (for 80 percent removal), or $25 to $65 (for 90 percent removal). Control of trihalomethanes in potable water by aeration, therefore, may result in increasing the total capital and operating costs for water treatment by 45 to 56 percent over total costs of $50 to $500 per million gallons for conventional treatment, as estimated by the American Water Works Association.

If the costs of removing 90 percent of trihalomethanes by aeration are compared to the costs of employing granular activated carbon adsorption, aeration apparently is more costly for treatment plants that have a capacity of 10 to 150 million gallons a day, and is significantly less costly only for plants at the 1 million gallons a day capacity. At lower efficiencies for trihalomethane removal, the cost of the aeration treatment is considerably reduced. Even though the ultimate residuals of chloroform are higher with a lower air-to-water ratio in aeration, the residuals (usually <100 µg/l) may still be acceptable and the actual costs will be lower than for carbon adsorption. Incremental costs for the control of trihalomethanes in potable water by various treatment options are summarized in Figure 7.1. Here again, the total incremental costs depend upon a plant's design capacity and upon the influent and effluent concentrations of trihalomethanes and their precursors. It is evident that control of trihalomethanes will generally result in significant increases in the total costs of water treatment. These increases are most apparent for smaller plants and for more stringent levels of control.

Ranges of the incremental and total costs for the treatment of water and trihalomethane control under various options are summarized in Table 7.5. Depending on the option and level of control that is selected, it is estimated that total costs will increase from a nominal amount to more than double the total costs of current conventional treatment. If the control level that is chosen is a mean concentration of 100 µg/l of trihalomethanes in finished water, then about 90 percent of existing plants could achieve this option with no special treatment and little additional cost. About 50 percent of existing water treatment plants currently meet a mean trihalomethane concentration of 20 µg/l, and many other plants could approach this level with improved conventional treatment. If the criterion for a control option were to be set below a trihalomethane concentration of 20 µg/l, however, special treatment would almost certainly be necessary and would require careful consideration of cost-benefit analyses for the lower criterion.

A preliminary economic impact analysis of proposed regulation of trihalomethanes in drinking water was prepared at the request of EPA (Temple, Barker, and Sloane, Inc. 1977). The basic premise of the authors is that approximately 52 percent of the U.S. population is served by about 390 water treatment systems in communities that have a population of 75,000 or more. They also estimate that about 86 (22 percent) of these facilities exceed a level of 100 µg total trihalomethanes present per liter of finished water. This compares to the national estimate that only about 10 percent of all water samples exceed this level. The differences in these data are apparently due to the greater use by large communities of surface water supplies, which may contain higher concentrations of precursors; smaller communities tend to make greater use of groundwater supplies. The data on trihalomethane concentrations that were used by Temple, Barker, and Sloan, Inc. were also maximized by storing the samples at ambient temperature for several days before analysis. In actual practice, consumption of treated water would take place in a shorter time span.

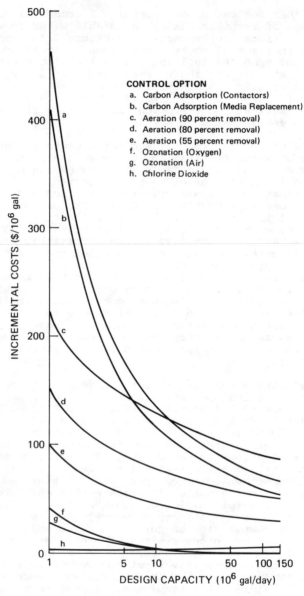

SOURCE: Data on Tables 7.2 and 7.4.

FIGURE 7.1 Comparison of incremental costs for disinfection (conversion from chlorine systems) and trihalomethane control by various methods.

TABLE 7.5 Ranges of Incremental and Total Costs for Various Levels of Trihalomethane Control in Municipal Water Supplies

Control Option	Mean Concentration of Trihalomethanes in Finished Water (μg/l)[a]	Incremental Cost for Control Option ($/10⁶) gal[b]	Total Cost of Delivered Water ($/10⁶) gal
Present Operation (90% of existing plants)	≤100	$—	$ 50-500[c]
Present Operation (50% of existing plants)	≤ 20	—	50-500[c]
Improved Conventional Treatment (all plants)	≤ 20	—	50-500[c]
Alternative Disinfectant	Background to ≤ 20	2- 40	52-540
Carbon Adsorption of Precursors	5-10	50-460	100-960
Carbon Adsorption of Formed Trihalomethanes	~0.1	50-460	
Aeration of Formed Trihalomethanes			
30:1 90%	10	85-220	135-720
16:1 80%	20	50-150	100-650
8:1 55%	45	30-100	80-600

SOURCE:
[a] Temple, Barker, and Sloane, Inc. (1977).
[b] Table 7.4.
[c] American Water Works Association (personal communication, Denver 1977).

The cost analysis by Temple, Barker, and Sloane, Inc. combines the assumptions of three major treatment options: (1) modifications of existing systems; (2) changing disinfectants; and (3) using a tertiary adsorbent. The first alternative involves minimal cost for systems that slightly exceed the proposed control level; the second alternative is somewhat more costly; the third alternative is the most complex and costly and it involves application of relatively unproven technology and uncertain costs. The economic implications of trihalomethane regulation at the 100 µg/l level were reported by the authors to be (in 1976 dollars): total capital expenditures of at least $154,000,000 from 1976 to 1981, and annual operating expenses of $26,000,000 by 1981. About 89 percent of the additional capital costs are incurred by 26 systems that may require the use of a tertiary adsorbent.

Several variables affect the sensitivity of the cost scenarios. The study by Temple, Barker, and Sloane, Inc. (1977) indicates that increasing the number of water treatment systems using adsorbents from 30 to 50 percent would increase capital expenditures on a national basis by up to $99,000,000, or 64 percent. Decreasing the permitted trihalomethane level to 50 µg/l would increase the number of systems that are affected to 141, and it would increase the total capital expenditures to $235,000,000. If the standard were set at 150 µg/l, 36 systems would be affected at a slightly reduced total capital expenditure of $103,000,000. In addition, if the lower boundary of population served were reduced from 75,000 to 10,000, almost six times as many systems (369) would require new treatment techniques, with the result that the nation's capital and operating expenses for water treatment would both increase by large amounts, the capital expenditures climbing from $154,000,000 to $319,000,000, and operation expenses going from $26,000,000 to $70,000,000 annually.

Regardless of the scenario that is selected, the control of trihalomethanes by tertiary adsorption methods represents significant costs. Modification of existing systems, such as reducing concentrations of chlorine, changing points of chlorine application, and pretreatment of water by coagulation and flocculation to remove the precursors of trihalomethanes are preferable alternatives when they can accomplish the task. Babcock and Singer (1977) have shown that pretreatment of raw waters by coagulation is an effective process for selective removal of precursors to chloroform. Humic and fulvic acids, which comprise over 80 percent by weight of all naturally-occurring organic material found in raw surface waters, were shown by these authors to be the major precursors of chloroform. Selective removal of both organic acids by coagulation with alum resulted in a decrease in concentrations of chloroform formed upon subsequent chlorination of the water, indicating that suppliers of public water where chloroform levels are high should consider coagulation as an option before chlorination is done.

In a related study (Young and Singer 1977), the feasibility of removing chloroform precursors by chemical coagulation, flocculation, and filtration using alum and cationic polyelectrolyte was evaluated at two municipal water treatment plants. Higher concentrations of nonvolatile total organic carbon (NVTOC) in the raw waters of both plants were generally followed by higher concentrations of chloroform after chlorination. The correlations were not exact, however, and could not be used for prediction, since NVTOC is a collective measure of all organic constituents and only a few of them are actual precursors of chloroform. Marked removals of precursors with alum, ferric chloride, and PURIFLOC^R C31 flocculant, a cationic polyelectrolyte, were demonstrated in laboratory studies. Prechlorination was terminated at one plant and disinfection was preceded by chemical coagulation and flocculation instead. A substantial reduction of 129 to 77 mg/l in chloroform production was achieved with an accompanying saving in chlorine use. Pretreatment to remove chloroform precursors before chlorination was recommended as an option to be evaluated at other plants if trihalomethane control becomes necessary.

OTHER NONFLUORINATED HALOMETHANES

A number of other nonfluorinated compounds are important industrial chemicals that are manufactured in large quantities. None has been identified as having significant adverse effects on human health at low levels of exposure, nor have quantitative estimates of human exposure been made.

The tropospheric burden of methyl chloride is relatively high, but this compound is thought to have a natural origin. None of the other compounds included in this study has been found to be significant as an environmental pollutant.

Methylene chloride and methyl chloride have been reported to be present at elevated levels in the indoor environment. Methylene chloride is a widely used solvent and has been found in indoor environments where the general public could be exposed at levels considerably above ambient levels but well below the exposure levels that are permitted in industry. Methyl chloride is a product of combustion reactions, including the smoking of cigarettes and the public is thereby exposed to isolated instances of elevated levels of this compound.

Since health and environmental concerns have not been identified and quantified for these compounds at or near anticipated exposure levels, work on control techniques and the costs of control have not been investigated in this report.

REFERENCES

Babcock, D.B. and P.C. Singer (1977) Chlorination and Coagulation of Humic and Fulvic Acids. Paper presented at the 97th Annual American Water Works Association Conference, May 8-13, Anaheim, California.

Burgess, R.A. (1976) Treatment, Recovery, and Occurrence of Halogenated Methanes: A Preliminary Literature Review. Graduate School of Public Health, University of Pittsburgh, June 28, 1976, Pittsburgh, Pennsylvania.

Love, O.T., Jr., J.K. Carswell, A.A. Stevens, and J.M. Symons (1975) Treatment of Drinking Water for Prevention and Removal of Halogenated Organic Compounds. Paper presented at the 95th Annual Conference, American Water Works Association, June 8-13, Minneapolis, Minnesota.

Morris, J.C. and G. McKay (1975) Formation of Halogenated Organics by Chlorination of Water Supplies. Office of Research and Development, U.S. Environmental Protection Agency. EPA-600/1-75-002; NTIS No. PB 241-511. Springfield, Va.: National Technical Information Service.

Morris, R.L. and L.G. Johnson (1976) Agricultural runoff as a source of halomethanes in drinking water. Journal of the American Water Works Association 68:492-4.

Rook, J.J. (1976) Haloforms in drinking water. Journal of the American Water Works Association 68:168-72.

Sawyer, C.M. (1976) Wastewater Disinfection: A State-of-the-Art Summary. Bulletin 89. Blacksburg, Va.: Virginia Water Resources Center, Virginia Polytechnic Institute and State University.

Stevens, A.A., C.J. Slocum, D.R. Seeger, and G.G. Robeck (1975) Chlorination of organics in drinking water. Pages 85-112, Proceedings of the Conference on the Environmental Impact of Water Chlorination, Oak Ridge, Tennessee, Oct. 22-24, 1975. NTIS Report CONF-751096. Springfield, Va.: National Technical Information Service.

Symons, J.M. (1976) Appendix 1, The Cost of Removing Chloroform and Other Trihalomethanes from Drinking Water Supplies. Interim Treatment Guide for the Control of Chloroform and Other Trihalomethanes. Water Supply Research Division, Municipal Environmental Research Laboratory, Office of Research and Development, U.S. Environmental Protection Agency. Cincinnati, Ohio: U.S. Environmental Protection Agency.

Temple, Barker, and Sloane, Inc. (1977) Economic Impact Analysis of a Trihalomethane Regulation for Drinking Water. Draft, prepared for the U.S. Environmental Protection Agency, Office of Water Supply.

TRW Systems Group (1975) Assessment of Industrial Hazardous Practices, Organic Chemicals, Pesticides, and Explosive Industries. Prepared for the U.S. Environmental Protection Agency; NTIS Report No. PB-251, 307. Springfield, Va.: National Technical Information Service.

U.S. Environmental Protection Agency (1976) Inventory of Public Water Supplies. Water Supply Division, EPA. Washington, D.C.: U.S. Environmental Protection Agency. (This serial publication is on computer tapes.)

Young, J.S., Jr. and P.C. Singer (1977) Chloroform Formation in Public Water Supplies: A Case Study. Paper presented at the 97th Annual American Water Works Association Conference, May 8-13, Anaheim, California.

CHAPTER 8

ECONOMIC ANALYSIS OF SELECTED METHODS FOR
REMOVING CHLOROFORM FROM DRINKING WATER[1]

The economic analysis in this chapter is primarily intended to demonstrate a methodology; the cost and benefit calculations are based on many assumptions and need refining and continual updating. New and more accurate data on the costs and benefits of regulating trihalomethanes are becoming available and the value of future calculations based on these new data should be greater.

ECONOMIC BENEFITS OF REDUCING HUMAN EXPOSURE TO TOXIC SUBSTANCES

A Perspective

From the viewpoint of economics, the central policy issue in controlling human exposure to any toxic substance is whether the benefits of reducing deaths, suffering, illness, and other losses outweigh the costs of controls. This involves identification of population exposure levels and a determination of when the costs of additional controls exceed the benefits of a further reduction in exposures. The uses and limits of benefit-cost analysis have been discussed in another NAS study, Decision Making for Regulating Chemicals in the Environment (NRC 1975).

In the judgment of many economists, estimating damages and benefits requires estimating the value of reducing the probability of an untimely death. But although these estimates are essential in making careful benefit-cost studies, they have been widely resisted. One reason for this is that the economic concept "cost of risk" has been confused by attempts to place a dollar value on human life. The idea of placing a dollar value on human life elicits justifiable moral and philosophical objections; however, this ethical dilemma must be confronted if realistic economic assessments are to be made. The expenditure of large amounts of money to prevent deaths from exposure to a substance of relatively low toxicity, for example, might be impracticable in a purely economic framework because it may take resources away from programs that could save more lives by reducing exposure to substances with greater toxicity. Choices of this kind are continually made by governments at all levels and by society, but the judgments about the value of life are often implicit rather than explicit in the decisions that are made. The National Academy of Sciences' study of decision making noted this fact and concluded that:

> "It may seem callous to trade dollars against human lives, but such valuation problems are unavoidable and are clearly implied by most of the important decisions the government makes...since valuation of noncommensurables is unavoidable, it is better for the decision maker to confront the choice of values openly and explicitly than to allow values to be hidden. In actual decisions the trade-offs are unlikely to be as brutally straightforward as dollars against lives, because there will be several incommensurable elements, not just one" (NRC 1975:49).

> "...benefit-cost analysis can provide the decision maker with a useful framework and language for describing and discussing trade-offs, noncommensurability, and uncertainty. It can help to clarify the existence of...value judgments concerning trade-offs" (NRC 1975:38).

Other objections to setting an economic value on the potential loss of human lives for benefit-cost studies stem from difficulties in making estimations and from the sensitivity of the results to whatever assumptions and procedures are followed. Continuing efforts are being made, however, to overcome these difficulties and to refine the concepts and procedures that are involved.

Concepts and Principles

The Discounted Value of an Individual's Production

The worth to society of the total production that an individual contributes to society over his or her lifetime is one concept that has been used to measure the value of preventing that person's death. Because future production and consumption generally are not considered as valuable in an economic sense as current production and consumption, future economic values are usually discounted to the present value by the market interest rate or some alternative discount figure.

The most serious shortcoming of this approach is that it defines the value of preventing death purely in terms of production and equates it to the market value of an individual's output rather than to other fundamental social values. This approach would, therefore, significantly undervalue or exclude the value of many human lives, including those of housewives, the elderly, the unemployed, and the underemployed, whose skills and capabilities either are not used in the marketplace or are not fully developed. Critics also contend that the purely economic contribution of workers is overvalued in this approach because the worker may consume a significant portion of the value of his or her output. In an attempt to counter this problem, an alternative concept of the value of a worker to society has been developed which establishes the value on the basis of the differential between discounted production and consumption data. Mishan (1971) has pointed out that this measurement technique undervalues the lives of the elderly and others who consume more than they currently produce. Furthermore, no allowance is made for the utility of life to an individual.

Extrapolations from Risk Premiums

Another concept that is now receiving attention from economists makes use of the amount paid to workers as a risk premium for engaging in dangerous occupations. The difference between the wages of a worker in a safe occupation and those of a worker in a hazardous occupation can be related to empirical measures (such as insurance rates) of the actual risk of death and injury. The results can then be extrapolated to provide an estimate of how workers actually compute the perceived risks of injury and, by extension, the value of avoiding death.

One shortcoming of this approach is that those who go into hazardous jobs may be less concerned about risks than the average person and this tends to give some downward bias to the value that is placed on avoiding death. Another problem with this approach is the long delay that is likely to occur between exposure and illness for persons who work with toxic substances, and the lack of information these workers may have about actual risks. Such factors would also tend to introduce a downward bias in the risk premium and, consequently, to estimates of the value of avoiding death.

Costs of Illness and Human Suffering

In determining the costs to society of illness and death as a result of exposure to toxic substances, the computations should include many costs in addition to loss of earnings. Among them are the direct costs of medical treatment and indirect costs from earnings foregone because of illness. Even if insurance covers such costs for the individual, much of the cost of medical care and unemployment is borne by society. Another cost that must be acccounted for is the distress and hardship imposed upon family and friends. Although it is difficult to assign an economic value to such costs, they can be computed by taking into account the length of an illness, the amount of physical and economic support that is required, and other quantifiable factors of these kinds.

The Pareto Improvement Principle

In order to derive rigorous, objective methods for conducting consistent benefit-cost evaluations of a program from which some people may benefit and others may not, economists have tried to develop methods that avoid interpersonal comparisons of what constitutes a benefit. The Pareto improvement principle is generally accepted for making such evaluations. Under this principle, an action is regarded as socially worthwhile if it can benefit some people without making anyone worse off. In a less rigorous formulation, the principle has been interpreted as requiring that an action must make some people sufficiently better off to enable them to compensate people who are initially made worse off to the extent that everyone ends up better off. One effect of the latter approach is to base the valuation of benefits upon the willingness to pay for them. Mishan (1971) describes the measurement of this value as "the equivalent variation," which is the maximum amount that individuals would be willing to pay to avoid added exposure to risk. An individual's "willingness" to pay, however, is limited by his or her ability to pay. To avoid this problem, Mishan has proposed another measure of value that he calls the "compensating variation"; this is defined as the minimum payment a person would demand in order to maintain his or her welfare at the previous level after exposure to increased health risks.

Empirical Estimates of Value of Reducing
the Probability of Death

Economists have recently made numerous quantitative estimates of the value of preventing a death, based on the concepts previously discussed, for use in benefit-cost studies of environmental management and other problems. The results vary widely, depending upon assumptions about price levels, discount rates, and other significant variables.

The critical role of the discount rate that is selected to derive data is illustrated by the high estimates for the

value of preventing death by economists who avoid
discounting altogether or who use low discount rates.
Conley (1976) concludes that, above some basic and critical
level of income, the value of death prevention is greater
than discounted earnings and that, in early and middle
adulthood, it is greater than discounted consumption.
Conley quotes estimates by Carlson (1963) of $200,000 to
$1,000,000 per life saved, based on U.S. expenditures for
defense, and on the compensation paid to pilots for flying
in high-risk situations. These estimates contain an upward
bias because they are based on the earnings of people who
are younger and have a greater life expectancy than the
average cancer victim and who also have a higher income than
the average person.

Schelling (1965) notes that the value of a particular
life is affected by personal values as well as by production
and consumption values. He estimates that a professional
worker would be willing to pay between ten and one hundred
times his or her annual earnings to save the life of a
family member. However, this applies to the cost of a
specific death, as contrasted with more widely used
statistical measures of a generalized risk of death.

Thaler and Rosen (1975) estimated the value of life by
analyzing labor market data and other information on the
earnings in different occupations, the job risks, and the
characteristics of people in various kinds of jobs. With a
risk of death in high-risk jobs of 0.001 per year (or one
death per thousand workers), Thaler and Rosen estimated that
the compensation required to make this risk acceptable is
about $200,000 ± $60,000 per individual per lifetime or per
1,000 workers per year. They did not consider the amount
that family and friends would be willing pay to save the
life of an individual in these estimates.

Bailey (1975) has suggested a number of refinements of
the Thaler-Rosen estimates, including technical corrections
for separation of the risk of death from the risk of injury,
risk avoidance, property income, tax, and third-party
effects. Bailey estimates that the value of reducing the
probability of a death by 100 percent is $286,000 when these
adjustments are made.

Cooper and Rice (1976) offer low per person estimates of
the cost of mortality from cancer: $39,808, discounting at
4 percent and $30,915, discounting at 6 percent. Although
they do not fully explain their methods in arriving at these
amounts the data are evidently based on an assumption of
death at relatively advanced ages. In addition to the cost
of mortality, these authors also estimated the costs of
morbidity, prevention, detection, treatment, rehabilitation,
training, and the capital investments that are involved in

these human activities. Dividing the Cooper and Rice data for the total cost of neoplasms in the United States in 1972 (more than $17 billion) by the number of deaths from this cause (nearly 353,000) produces an average cost of $49,226 for each death. If the adjustments for various indirect costs, as developed by Bailey (1975) were used in these calculations, the average cost of a death from neoplasm would be $210,814.

An additional adjustment could be made in these estimates to account for the lag between exposure to a carcinogen and any resulting death, which averages about 20 years. Appropriate discount rates for future years can be obtained by subtracting the rate of inflation (about 7 percent) and the rate of growth in productivity (2 percent) from currently inflated market interest rates (about 10 percent) in order to get an adjusted real discount rate of 1 percent. Discounting the above estimates for 20 years by 1 percent yields, a figure for the average cost of a death from neoplasm of $172,770. Discounting more heavily by using a 4 percent rate yields an estimate of $96,213. While general agreement exists concerning the economic validity of discounting values of future goods and services, by appropriately adjusted market interest rates, the value of future lives is an ethical, not an economic issue, and therefore the decision on what discount, if any, to apply to future lives should be an ethical or political decision.

Conclusions

Depending on the methodology that is used to compute costs, from these examples, the most reasonable estimates of the per capita value associated with reducing the probability of death by 100 percent range from $100,000 to $1,000,000. Thus, for example, where there would be a reduction in incremental risk of 1 percent over an average lifetime, this would correspond to a $1,000 to $10,000 per capita value to society. Because of the sensitivity of such estimates to the assumptions that are involved, the estimating process should be regarded as an illustration of the concepts and methodologies involved in determining a statistical measure of a generalized risk of death, not as precise, scientific calculations of the economic value of human life. The inherent imprecision in the calculations is also the reason that the benefit-cost analyses later in this chapter use a range of values of $100,000 and $1,000,000 as the per person per population benefit for preventing death.

COSTS OF REMOVING CHLOROFORM AND OTHER TRIHALOMETHANES
FROM DRINKING WATER

The cost side of the benefit-cost equation that is used in the control of toxic substances must be calculated for each specific control technique because the costs per person benefited may vary greatly among the available control options. This Panel selected two promising systems for removal of chloroform and other trihalomethanes from drinking water supplies to demonstrate the calculation of costs in a control program that is likely to have wide application. One of these systems is the granular activated carbon (GAC) process that is applied to water before the introduction of chlorine which can act on precursors to generate chloroform. According to Clark et al. (1976) and Symons et al. (1976), the GAC process is effective in removing not only chloroform and other trihalomethanes, but also other potentially toxic substances. The second system is aeration treatment in which a 30:1 air to water ratio is used to achieve a 90 percent removal of chloroform and other compounds.

The costs of the two systems are compared in Tables 8.1 and 8.2. These tables assume a cost per capita based on a daily per capita municipal water treatment volume of 143 gallons (Porges 1957). A more recent study indicates that 166 gallons per capita per day is the national average for U.S. public water supplies (Murray and Reeves 1972). Data from the 99 largest municipal water treatment plants in the United States (U.S. EPA 1976) indicate that 200 gallons per day is the national per capita figure. The volume per capita varies considerably throughout the United States, with regional values of less than 100 and higher than 300 gallons/capita/day. As this volume varies, so would the annual per capita cost for a given size facility. Thus, as the costs vary, the benefit-cost balance point obviously varies as well.

For small plants, the aeration treatment system is less costly than the GAC media replacement system, while the reverse is true for large plants. For example, total costs of the aeration system in plants with a capacity of one million gallons per day are about half those of the GAC system; however, in plants with a capacity of 150 million gallons a day (MGD), total costs of treatment by the GAC process are only about 60 percent of what the aeration process would cost. The main reason for this situation is that capital costs for the GAC system fall sharply with increases in plant size. In terms of per capita costs, shown on Table 8.2, the annual cost of the GAC system falls from $21.40 in treatment plants with a capacity of one million gallons a day to about $2.66 for plants with a capacity of 150 million gallons a day. Per capita costs for

TABLE 8.1 Costs of Two Chloroform Removal Systems (¢/1,000 gallons)

Size of Facility (10⁶ gal/day)	Capital Cost		Operating and Maintenance Cost		Total Cost	
	GAC System	Aeration System (90% removal)	GAC System	Aeration System (90% removal)	GAC System	Aeration System (90% removal)
1	21.5¢	11.6¢	21.5¢	10.7¢	41.0¢	22.3¢
5	5.0	7.2	10.5	7.4	15.5	14.6
10	3.5	6.0	8.2	6.7	11.7	12.8
100	1.5	3.6	4.5	5.4	6.0	9.0
150	1.1	3.2	4.0	5.3	5.1	8.5

SOURCE: Clark et al. (1976) and panel estimates.

TABLE 8.2 Annual Per Capita Costs of Two Chloroform Removal Systems

Size of Facility (10⁶ gal/day)	Estimated Population Served	GAC Media Replacement System		Aeration Treatment System (90% removal)	
		Cost/ 1,000 gal	Annual Cost per capita[a]	Cost/ 1,000 gal	Annual Cost per capita[a]
1	2,300	$0.410	$21.40	$0.223	$11.64
5	11,600	0.155	8.09	0.146	7.67
10	23,300	0.117	6.11	0.128	6.68
100	233,000	0.060	3.13	0.090	4.70
150	348,000	0.051	2.66	0.085	4.44

[a]Assuming 52,195 gallons of water treated per year per capita (Porges 1957).

SOURCE: Clark et al. (1976) for GAC process; Panel Estimates for aeration process.

the aeration treatment system fall from $11.64 to $4.44 in plants with these capacities.

In areas without treatment plants, and for persons using individual wells, the cost per person of reducing chloroform in drinking water could be prohibitively high, unless the systems were so well designed and maintained that organic precursors could be excluded. The use of bottled water would be an expensive alternative for people in these areas.

Although some water consumers live on farms and in other areas where they depend on individual wells for their water, it is useful for statistical purposes to estimate what the costs of drinking water treatment systems would be for the U.S. population as a whole. At the greatest possible economies of scale, and at annual per capita costs of $2.66 for the GAC system and $4.44 for the aeration system, the minimum total U.S. costs for these respective systems would be about $572 million and $955 million. Actual nationwide costs would be much greater, however, because of the diseconomies of scale in small communities. Whether these costs are justified depends largely upon the estimated value of benefits.

BENEFITS OF REMOVING CHLOROFORM FROM DRINKING WATER SUPPLIES

The subsequent calculations assume that the most significant effect of human exposure to chloroform in drinking water is cancer and that all of these cancers result in death; effects other than cancer mortality are negligible. The benefits of reducing human exposure to chloroform in drinking water can be estimated by multiplying data on lifetime risk of cancer by the economic value of reducing the risk of death from cancer in a population. The benefits also may be calculated by multiplying the daily per capita uptakes of chloroform by the risk of a cancer death over an average lifetime from a given daily dose of the carcinogen by the economic value of reducing the risk of a cancer death. The most probable lifetime cancer risk (r) for a 70 kg adult was calculated in Chapter 6 to be 2×10^{-7} for an oral dose of 1 µg chloroform/day. The upper limit risk, or upper 95 percent confidence bound, increases this estimate by a factor of 5.

The basic formula applied by the Panel in the derivation of annual per capita benefits is

$$B = (r_1/60) \times v$$

where B is per capita annual benefits in dollars from reduction of cancer assuming 100 percent chloroform removal,

r_1 is the per capita lifetime risk estimate under specified uptake conditions, v is the assumed economic value in dollars of preventing a cancer death, and 60 is the average life expectancy.

Because of the lack of information on variation in risk throughout lifetime, the risk estimates assume constant risk for each age throughout lifetime. The lifetime risk estimate was divided by 60 in order to obtain annual risk without taking account of carcinogenic lag and other age-specific data. If additional age-specific data become available, revised benefit estimates related to age could be computed. In fact, the average period of lifetime quantitative exposure to carcinogens in drinking water in a population is related not only to life expectancy, but also habits in use of that drinking water. Even for an individual, such habits will vary throughout a lifetime. In the United States, the average life expectancy at birth was 68 years for males and 76 years for females in 1974 (WHO 1977). Among 39 nations reported for the period 1973-1975, the comparable figure for males varied from 61 to 73 years. If 70 years (perhaps a better average life expectancy) were used in our calculations, it would reduce the average annual death rate and hence the annual per capita benefit in the benefit-cost analysis by about 17 percent. This would result in making the removal of chloroform from drinking water that much less cost effective.

The derivation of r_1 is

$$r_1 = C \times V \times r$$

where C is the initial concentration of chloroform in water (µg/liter) before treatment, V is the daily per capita intake of water (liters/day), and r is 2×10^{-7} for most probable lifetime risk and $2 \times 10^{-7} \times 5$ for upper limit lifetime risk at an oral dose of 1 µg chloroform/day.

Most data for each element in the above formulas are from Chapter 6. Data for the initial concentrations of chloroform in finished drinking water are current in the available literature through 1976 and are shown in Table 6.6. This study uses 0.1, 21, and 366 µg/l for the minimum, median, and maximum concentrations of chloroform found in finished drinking water in the United States. More recent data on trihalomethane concentrations in water supplies reported in an unpublished EPA study (National Organics Monitoring Study, 1977, Cincinnati, Ohio), have not been evaluated in this report. Data for measured daily water intakes are shown in Tables 6.1 and 6.2: 1.0, 1.95, and 3.7 liters/day are used, for minimum, reference man, and maximum adult fluid intakes. Per capita lifetime risk estimates (r_1), corresponding to various uptake conditions (concentration and intake), are shown in Table 6.23. The calculated upper 95 percent confidence bound on these risk calculations would increase these estimates by a factor of 5.

As an example of the application of the above basic benefit formula, suppose a situation where there is a median concentration of chloroform (C = 21 µg/liter), an intake of fluid at the reference man level (V = 1.95 liter per day), a given value of preventing a cancer death at $300,000 (v = $300,000), and an upper limit risk. Then the annual per capita benefit, as shown in Table 8.4, is

$$B = (r_1/60) \times v$$

if $r_1 = C \times V \times r$, and r is $5(2 \times 10^{-7})$ for upper limit risk,

then $B = \dfrac{C \times V \times r}{60} \times V$

$= \dfrac{21 \times 1.95 \times 5 \times 2 \times 10^{-7}}{60} \times \$300,000$

$= \$0.20.$

With 100 percent removal of chloroform from drinking water, the estimated per capita annual benefits for most probable cancer risk and upper limit risk are shown in Table 8.3 and Table 8.4, respectively. The last columns of these tables show the estimated numbers of deaths per year from cancer for the U.S. population under the given conditions of risk.

Under conditions of minimum chloroform concentration (see the first three rows in Table 8.3 and Table 8.4) the benefit never rises to a measurable level under conditions of either most probable or upper limit risk. If the case were one of a median concentration of chloroform (21 µg/liter in U.S. municipal water supplies), however, and daily per capita fluid intake for reference man (1.95 liters), with the most probable risk factor and a $300,000 value of preventing death, the annual per capita benefit from treatment would be $0.04, as shown in Table 8.3. If all conditions are at the highest level--there is a maximum chloroform concentration (366 µg/liter), maximum fluid intake (3.7 liter/day), and the estimated value of preventing a cancer death is $1,000,000,--the annual benefit per capita for the upper limit risk is estimated to be $22.50, as shown on Table 8.4.

Estimates of the number of cancer deaths that result annually from various degrees of chloroform uptake are zero under minimum concentration-minimum intake conditions, 29 under typical conditions (median concentration and reference man intake), and 968 with conditions of maximum concentration and maximum intake, given a situation of most

TABLE 8.3 Estimated Per Capita Annual Benefits for 100% Removal of Chloroform from Drinking Water (Assuming Most Probable Risk of Death from Cancer)

Uptake Conditions	Lifetime Risk Estimates	Assumed Economic Value of Reducing the Risk of a Cancer Death						Estimated Annual U.S. Cancer Deaths[a]
		$100,000	$200,000	$300,000	$400,000	$500,000	$1,000,000	
Minimum Concentration								
Min. Intake	2×10^{-8}	$0	$0	$0	$0	$0	$0	0.07
Ref. Man Intake	3.9×10^{-8}	0	0	0	0	0	0	0.14
Max. Intake	7.4×10^{-8}	0	0	0	0	0	0	0.27
Median Concentration								
Min. Intake	4.2×10^{-6}	0	0.01	0.02	0.03	0.04	0.07	15
Ref. Man Intake	8.1×10^{-6}	0.01	0.03	0.04	0.05	0.08	0.14	29
Max. Intake	1.6×10^{-5}	0.03	0.05	0.08	0.11	0.13	0.27	57
Maximum Concentration								
Min. Intake	7.3×10^{-5}	0.12	0.24	0.37	0.49	0.60	1.22	261
Ref. Man Intake	1.4×10^{-4}	0.23	0.47	0.70	0.93	1.17	2.33	501
Max. Intake	2.7×10^{-4}	0.45	0.90	1.35	1.80	2.25	4.50	967

[a] Calculated by multiplying estimated lifetime risk x population at risk x 1/average lifetime, or estimated lifetime risk x 215 x 10^6 x 1/60.

TABLE 8.4 Estimated Per Capita Annual Benefits for 100% Removal of Chloroform from Drinking Water (Assuming Upper Limit Risk of Death from Cancer)

Uptake Conditions	Lifetime Risk Estimates	Assumed Economic Value of Reducing the Risk of a Cancer Death						Estimated Annual U.S. Cancer Deaths[a]
		$100,000	$200,000	$300,000	$400,000	$500,000	$1,000,000	
Minimum Concentration								
Min. Intake	$2 \times 10^{-8} \times 5$	$0	$0	$0	$0	$0	$0	0.35
Ref. Man Intake	$3.9 \times 10^{-8} \times 5$	0	0	0	0	0	0	0.70
Max. Intake	$7.4 \times 10^{-8} \times 5$	0	0	0	0	0	0	1.35
Median Concentration								
Min. Intake	$4.2 \times 10^{-6} \times 5$	0.04	0.07	0.11	0.14	0.18	0.35	75
Ref. Man Intake	$8.1 \times 10^{-6} \times 5$	0.07	0.14	0.20	0.27	0.34	0.68	145
Max. Intake	$1.6 \times 10^{-5} \times 5$	0.13	0.27	0.40	0.53	0.66	1.33	286
Maximum Concentration								
Min. Intake	$7.3 \times 10^{-5} \times 5$	0.61	1.22	1.83	2.43	3.04	6.08	1,307
Ref. Man Intake	$1.4 \times 10^{-4} \times 5$	1.17	2.33	3.50	4.67	5.83	11.67	2,508
Max. Intake	$2.7 \times 10^{-4} \times 5$	2.25	4.50	6.75	9.00	11.25	22.50	4,837

[a] Calculated by multiplying estimated lifetime risk x population at risk x 1/average lifetime, or estimated lifetime risk x 215 x 10^6 x 1/60.

probable risk. If upper limit risk is assumed, these same data are estimated at 0, 145, and 4,838, respectively, as shown in the last column of Table 8.4.

As was true in the calculation of costs, a number of assumptions enter into the calculation of benefits. The variables used in making estimates were treated as deterministic, and any unpredictability or variability in the data was ignored. For example, the risk estimate (r), which was calculated in Chapter 6 by making extrapolations from tests on laboratory animals, was assumed to be a linear function. Additionally, it was recognized that there may be problems of synergism with other substances, however, this was not addressed. Furthermore, the benefit estimates in Tables 8.3 and 8.4 were based on removing only chloroform from drinking water. Carbon tetrachloride is also found in drinking water and can also be reduced by the GAC process. Data on its concentrations and related risks are given in Chapter 6. The benefits from a reduction of carbon tetrachloride in drinking water are not measurably significant, however, because of the low concentration of this compound; as result, these benefits would make little or no difference in the estimates of benefits resulting from treatment of water supplies. It is possible that if the risk factors for a number of substances such as pesticides, were also taken into account in the calculation of benefit-cost equations, the benefits of their removal would achieve greater statistical significance, but insufficient data are available at this time to calculate the dollar benefits of their removal from drinking water. It should also be noted that the emphasis in this study on benefits and costs in water treatment does not take account of such benefits as improvements in the taste of water, or reduction in costs incurred in the use of household filtration processes, or bottled water. Finally, since the aeration treatment process removes only 90 percent of chloroform and carbon tetrachloride, compared to 100 percent removal with the GAC process, the benefit figures in Tables 8.3 and 8.4 should be reduced by 10 percent when applied to the aeration process. This is not shown in the tables because the adjustment did not noticeably change the results for any case presented in these tables.

BENEFIT-COST ANALYSES FOR GAC AND AERATION SYSTEMS

The data on benefits in Tables 8.3 and 8.4 can be compared with per capita costs in Table 8.2 to determine whether or not a *prima facie* economic case exists for application of either the granular activated carbon (GAC) system or the 30:1, air:water aeration system for the treatment of water supplies.

Table 8.2 indicates that the minimum annual cost per capita for 100 percent removal of chloroform is $2.66. This is for treatment plants with a 150 million gallons/day capacity that use the GAC system. The value of benefits for the most probable risk from chloroform in drinking water rises above this minimum cost level only in the case of maximum fluid intake and maximum initial chloroform concentration and with the value of preventing death at $1,000,000, which is the highest assumed value. If an upper limit of risk (95 percent confidence estimate) is assumed, benefits exceed the costs of 150 million gallons/day GAC plants in more cases, these cases include: minimum intake-maximum concentration when the prevention of death is valued at $500,000 or more; when there is reference man intake-maximum concentration with prevention of death valued at $300,000 or more; and when there is maximum intake-maximum concentration, if preventing a death is valued at $200,000 or more.

Because treatment costs using the aeration system are higher than in the GAC system, except for plants with a capacity of 5 million or less gallons a day, the cases in which the aeration system offers greater benefits than costs are fewer than for the GAC system. As was the case in the GAC system, the value of benefits is greater than costs for 150 million gallons/day plants when the most probable risks are assumed only in the extreme case of maximum concentration-maximum intake with prevention of death valued at $1,000,000. When the upper limit of risk and maximum initial concentration of chloroform are assumed, the aeration system in 150 million gallons/day treatment plants is economically justified in such instances as: minimum fluid intake with the value of preventing a death set at $1,000,000; reference man intake when prevention of death is valued at $400,000 or more; and maximum intake for all values that are assigned to the prevention of death except for the lowest value of $100,000.

In either the aeration or the GAC system, treatment is also economically justified even for plants with 5 million gallons/day capacity when there is maximum concentration-reference man intake with the value of preventing death at $1,000,000, and maximum concentration-maximum intake and the prevention of a death is valued at $400,000 or more.

Some aspects of the benefit-cost relationships discussed above are shown in Figures 8.1 and 8.2. Figure 8.1 is scaled for the most probable risk estimates, and Figure 8.2 for upper limit risk estimates and both figures use selected data from Tables 8.2, 8.3, and 8.4. The figures compare the annual per capita costs of 5 MGD and 150 MGD capacity plants for each of the two treatment systems, and show the per capita annual benefit from 100 percent removal of chloroform

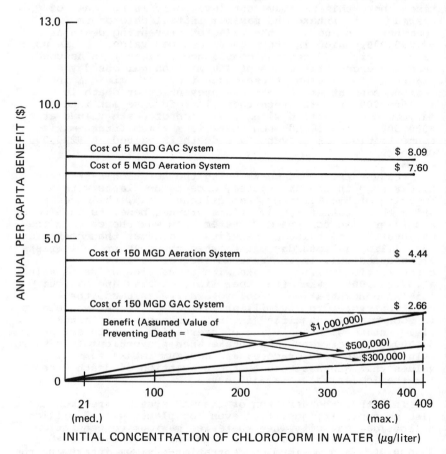

FIGURE 8.1 Annual per capita benefits (assume reference man intake by 70 kg adult) and costs of removing chloroform from drinking water as a function of initial chloroform concentration (*assumes most probable risk*).

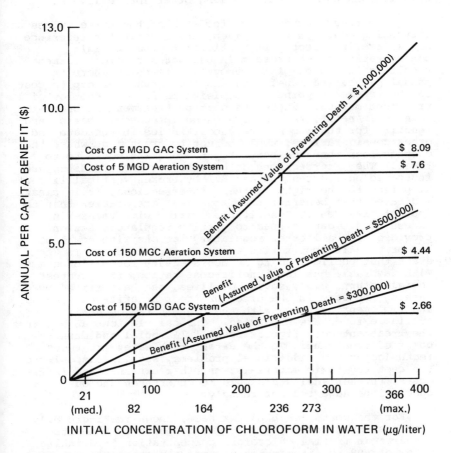

FIGURE 8.2 Annual per capita benefits (assume reference man intake by 70 kg adult) and costs of removing chloroform from drinking water as a function of initial chloroform concentration (*assumes upper limit risk*).

against the initial concentration of chloroform for three values of preventing death: $300,000, $500,000, and $1,000,000. The benefit curves assume a 1.95 liters per day intake by a 70 kg man (i.e., reference man). Similar calculations can be made assuming other intake levels.

These figures show that, for each of the three values of preventing death that are given, the benefit for reference man intake is directly proportional to the initial concentration of chloroform in the water before treatment resulting in its complete removal. The point where the annual per capita benefit equals the annual per capita cost is the reference point for decision making on adoption of a treatment method. Where the cost of treatment is greater than the benefit, the conclusion on an economic basis, and ignoring for the moment the uncertainties in the data and other considerations, would be not to treat and, where the cost is less than the benefit, the judgment would be to treat. The intersections of benefit and cost findings are indicated on Figures 8.1 and 8.2 by broken vertical lines. At points to the right of these intersections, it is clearly indicated that benefits from treatment are greater than the cost of treatment. Benefit-cost ratios with values in excess of 1.0 can be used to justify regulatory action provided that all other regulatory actions with equal or higher benefit-cost ratios are also implemented. In the event that public funding is not adequate for all programs with favorable but low benefit-cost ratios, the greatest social return, e.g., saving of lives, can be obtained by ranking projects in the order of their benefit-cost ratios and implementing those with the highest ratios until available funds are exhausted. Our calculations assume that the treatment cost figures cover full social and market costs of treatment. To the extent that water treatment technology causes additional problems, such as disposing of the captured toxic materials, or other external and indirect costs, further analysis should be undertaken and our conclusions adjusted accordingly.

For the most probable risk case, shown in Figure 8.1, the break-even level for costs as compared to benefits requires an initial chloroform concentration in drinking water of 409 µg/liter (which exceeds the maximum level of 366 µg/liter in drinking water that was found in this study), even when a value of $1,000,000 is assigned to reducing the risk of death. This Panel has concluded that, given the most probable risk, removal of chloroform from drinking water using either of the treatment systems now available cannot be justified on economic grounds because the per capita cost exceeds the value of risk reduction under the conditions analyzed in this report.

Figure 8.2, which assumes the upper limit risk, suggests that if preventing death is valued at $1,000,000, in large communities (population of about 348,000) where the full economies of scale in water treatment can be captured at 150 million gallons/day, there is economic justification for installation of a GAC system if the initial concentration of chloroform exceeds 82 µg/liter. Even in small communities, where all treatment plants are expensive to operate but where the aeration process is less costly at a capacity of 5 MGD, economic justification exists for treatment if the initial concentration of chloroform exceeds 236 µg/liter. If the prevention of death is valued at $500,000, a 150 million gallons/day GAC system treatment is justified if the initial concentration of chloroform is more than 164 µg/liter. Even if prevention of a death is valued as low as $300,000, economic justification for GAC treatment (but not for aeration treatment) exists if initial concentrations of chloroform exceed 273 µg/liter for plants with a capacity of 150 million gallons/day.

The task of regulating chloroform content in drinking water is further complicated by present knowledge about treatment processes. If treatment processes were available that are flexible and permit the level of chloroform removal to increase with cost per treatment, computations of optimal or acceptable levels of chloroform in drinking water could be carried out and standards could be recommended. The information currently available indicates that the GAC system operates efficiently only at the level of nearly 100 percent removal of chloroform precursors, while the aeration process is more flexible in its uses. Rather than attempt to recommend appropriate levels of chloroform regulation, this study has tried to describe a methodology that would help to provide the cost-benefit information to decision makers in the most explicit form. <u>Using the best currently available data, it has been possible to use this method to indicate the initial levels of chloroform concentration that would justify 90 percent removal through aeration or 100 percent removal by the GAC process, based only on economic considerations.</u> If regulations are promulgated for removal of chloroform from drinking water in communities with small treatment plants and high initial concentrations, further calculation should be made in order to determine optimal levels of removal using the aeration process, since this process is flexible.

In addition to the problem of the overall economic justification for incurring costs to reduce the risk from hazardous substances in drinking water, there are also important questions about the distribution of benefits and costs that would arise from the introduction of treatment processes and from different approaches to the administration and financing of these processes. In

general, the major alternative control strategies include
the use of regulation standards to prescribe certain levels
of environmental quality, subsidies to reward those who take
specific actions to reduce pollution, and charges or taxes
levied against persons who pollute the environment. Any one
of these strategies can be effective, but with different
economic effects. The most widely used instruments of
environmental management in the United States have been the
regulation of standards (in water quality) and federal
subsidization of local sewage plant construction. These
strategies have been more successful, for example, than the
imposition by the federal government of charges on the
emission of hazardous substances by private firms. The fact
that most water plants are operated by municipal governments
may raise major legal and administrative obstacles to the
imposition of federal charges on chloroform in drinking
water. Therefore, where chloroform removal from drinking
water can be justified on economic or other grounds, serious
consideration should be given to federal programs that
emphasize the importance of regulating chloroform and that
offer financial assistance to local governments in achieving
this objective.

CONCLUSIONS

Panel calculations, which are based on the best
available data, show that very high concentrations of
chloroform in drinking water are associated with enough risk
of cancer to justify the costs, on economic grounds alone,
of treatment processes for removal of this compound. The
potential magnitude of the problem is even greater if
allowance is made for the upper limit of risk. Furthermore,
justification for treatment rises with the value imputed to
avoiding a death. However, the current cost of treatment to
remove chloroform from drinking water is sufficiently high
that the economic justification for removing chloroform from
drinking water in the United States, assuming the most
probable risk, exists only in those cases where maximum
initial concentrations of chloroform are found in drinking
water, there is maximum fluid intake, and the risk of death
is valued at $1,000,000 or more. Using a more typical and
more statistically justifiable value of reducing the risk of
death, i.e., $300,000 (see earlier discussion of Bailey's
refinements of the Thaler-Rosen estimates in this chapter,
the section on Empirical Estimates of Value of Reducing the
Probability of Death), the high cost of removing chloroform
alone cannot be justified on economic grounds for the most
probable risk conditions, even when there are maximum
concentrations and intake.

If a safety factor is allowed by assuming an upper limit
risk of five times that of the most probable risk, and if a

death avoided is conservatively valued at $300,000, the cost of chloroform removal can be justified for large communities where initial concentrations of chloroform are high (273 μg/liter), even given the average levels of intake for reference man. If the value of preventing a death is assumed to be $500,000, chloroform removal is justified for maximum concentrations even when intakes are at minimum levels. The above estimates are contingent, however, upon the assumption that the most efficient treatment processes are being used and that they are operating at maximum economies of scale. For smaller communities, where not all economies of scale in water treatment can be attained, the benefits of treatment remain the same; because the costs of treatment rise, however, there is a lessened economic justification for treatment of drinking water. For small communities that are limited to treatment processes of five million gallons per day, it is necessary to assume a value of $1,000,000 for a death avoided for the level of intake by reference man to justify the cost of removing chloroform from drinking water supplies.

These conclusions about the economic justification of chloroform removal are, as noted earlier, very sensitive to the assumptions that are made. Removal of chloroform is difficult to justify under current typical conditions of low concentrations in drinking water, most probable risk, an economically justifiable value of preventing death, and current high costs for chloroform removal. The benefits of chloroform removal are sharply increased, however, by including the safety margin of upper limit risks, high initial concentrations, and high values of avoiding death. Since the two major processes studied here for chloroform removal from drinking water also remove carbon tetrachloride, insecticides, and probably a limited variety of other hazardous substances as well, the total benefits to be derived from drinking water treatment are greater than those for the removal of chloroform alone.

It is evident that additional research on the possible hazards of other substances in drinking water and on the development of less costly methods for removing hazardous substances from drinking water might greatly help to reduce death and illnesses in the United States that result from carcinogens in drinking water. Given the current knowledge concerning this problem, economic justification can still be found for the application of current treatment technologies for the removal of chloroform where it exists in high concentrations in large communities in the United States and serious consideration should be given to the establishment of a federal program to help bring this about.

NOTE

1. The method used in this chapter to estimate benefits required estimating the value of reducing the probability of a human death. It was not the unanimous opinion of the Panel to use this concept. In particular, Dr. David Hoel disagrees with the concept, but agreed to its inclusion in the report with his exception noted.

REFERENCES

Bailey, M.J. (1975) Benefits of Safety. College Park, Maryland: University of Maryland. (Unpublished manuscript)

Carlson, J.W. (1963) Valuation of Life Saving. Unpublished doctoral dissertation. Cambridge, Mass.: Harvard University.

Clark, R.M., D.L. Guttman, J.L. Crawford, and J.A. Machisko (1976) The Cost of Removing Chloroform and Other Trihalomethanes from Drinking Water Supplies. Municipal Environmental Research Laboratory, Office of Research and Development. EPA 600/1-77-008. Cincinnati, Ohio: U.S. Environmental Protection Agency.

Conley, B.C. (1976) The value of human life in the demand for safety. American Economic Review 66:46-55.

Cooper, B.S. and D.P. Rice (1976) The economic cost of illness revisited. Social Security Bulletin 39:21-36.

Mishan, E.J. (1971) Evaluation of life and limb: A theoretical approach. Journal of Political Economy 79:697-705.

Murray, C.R. and E.B. Reeves (1972) Estimated Use of Water in the U.S., 1970. U.S. Geological Survey. Washington, D.C: U.S. Department of Interior.

National Research Council (1975) Decision Making for Regulating Chemicals in the Environment. Committee on Principles of Decision Making for Regulating Chemicals in the Environment. Washington, D.C.: National Academy of Sciences.

Porges, R. (1957) Factors influencing per capita water consumption. Water and Sewage Works 104:199-204.

Schelling, T.C. (1965) The life you save may be your own. In Problems in Public Expenditure, S.B. Chase, Jr., editor. Washington, D.C.: The Brookings Institution.

Symons, J.M. (1976) Interim Treatment Guide for the Control of Chloroform and Other Trihalomethanes. Water Supply Research Division, Municipal Environmental Research

Laboratory, Office of Research and Development, U.S. Environmental Protection Agency. Cincinnati, Ohio: U.S. Environmental Protection Agency.

Thaler, R. and S. Rosen (1975) The value of saving a life: Evidence from the labor market. In Household Production and Consumption, Studies in Income and Wealth, Volume 40, edited by N.E. Terleckyj. New York: National Bureau of Economic Research.

U.S. Environmental Protection Agency (1976) Inventory of Public Water Supplies. Washington, D.C.: Water Supply Division, U.S. Environmental Protection Agency. (A serial publication on computer tapes.)

World Health Organization (1977) Vital Statistics and Cause of Death. Vol. 1, World Health Statistic's Annual. Geneva: World Health Organization.

APPENDIX A

PHYSICOCHEMICAL PROPERTIES OF THE
NONFLUORINATED HALOMETHANES

CHLOROMETHANES

Methyl Chloride (CH_3Cl)

General Description. A colorless, almost odorless gas which can be compressed to a colorless liquid that has a faint, nonirritating, ethereal odor and a sweet taste. It is soluble in water and ethyl alcohol and is miscible in all proportions with diethyl ether, acetone, benzene, chloroform, and acetic acid. (Other names: chloromethane, monochloromethane.)

		Reference[†]
Molecular weight:	50.49	K,L
Melting point:	-97.7°C	K,H
Boiling point:	-24.2°C	K
Density (see Figure 3.2):	$0.9159(^{20}_{4})$	K
Vapor specific gravity:	1.78 g/l (air = 1.0)	

Vapor pressure:
(see Figure 3.1)

Pres. (mm Hg)	Temp. (°C)	
760	-24.2	E
1,520	- 6.4	K
3,756	+20.0	E
3,800	+22.0	K
4,311.4	+25.0	C

Solubility in water:

ppm (mg/l)	Temp. (°C)	
7,600	0	A
~4,600-8,000	20	A,E
~4,800-7,000	25	A,H
7,300	30	A

Partition coefficient:
($*C_s$ = saturation concentration)

Temp. (°C)	$*C_s$ air/C_s water (w/v per w/v)		
	Calc'd	Found	
20	0.33	0.30	A
20	--	0.30	E

Evaporation rates:
(half-life)

	50% Depletion		90% Depletion
Reference	A	B	B
Time, min.	27.6	27	91

Conversion factors:
(25°C, 1 atm in air)

1 ppm (v/v) = 0.0021 mg/l
1 mg/l = 484.2 ppm (v/v)
1 ppm (v/v) = 2.065 mg/m³
1 mg/m³ = 0.484 ppm (v/v)

[†] Key to lettered references appears at end of appendix.

Methylene Chloride (CH_2Cl_2)

General Description. A colorless, highly volatile liquid with a penetrating, pleasant, ethereal odor (odor threshold about 214 ppm). It is nonflammable and normally nonexplosive and it is miscible in all proportions with ethyl alcohol and diethyl ether, but is only slightly soluble in water. (Other names: methane dichloride, dichloromethane, methylene bichloride, methylene dichloride.)

		Reference
Molecular weight:	84.93	
Melting point:	-96,7°C	K,H,J
Boiling point:	40.0°C	K,F
Liquid density: (see Figure 3.2)	1.326 g/ml (20°C)	K,C,J
Vapor specific gravity:	2.93 g/l (air = 1.0)	H

Vapor pressure: (see Figure 3.1)

Pres. (mm Hg)	Temp. (°C)	
147.4	0	K,H
∼230	10	A,H
∼350	20	A,H,E
435.9	25	C
511.4	30	H
600	35	H

Solubility in water:

ppm (mg/l)	g/100g water	Temp. (°C)	
23,100	2.36	0	H
20,800	2.12	10	H
19,600	2.00	20	H
20,000	--	25	A
19,000	1.97	30	H

Partition coefficient: (*C_s = saturation concentration)

Temp. (°C)	*C_s air/C_s water (w/v per w/v) Calc'd	Found	
20	--	0.12	E
20	0.085	0.12	A
25	0.10	0.11	A

Evaporation rates: (half-life)

	50% Depletion				90% Depletion
Reference	A	A	A	B	B
Avg. time, min.	18.4	25.2	20.7	21	69

Conversion factors: (25°C, 1 atm in air)

1 ppm (v/v) = 0.0035 mg/l
1 mg/l = 287.9 ppm (v/v)
1 ppm (v/v) = 3.474 mg/m^3
1 mg/m^3 = 0.288 ppm (v/v)

Chloroform ($CHCl_3$)

General Description. A clear, colorless, nonflammable liquid with a characteristic heavy, pleasant, ethereal odor and a burning sweet taste. (Odor threshold about 676 ppm.) It is miscible in all proportions with ethyl alcohol, diethyl ether, benzene, and naptha; it is soluble in acetone, and slightly soluble in water. (Other names: trichloromethane, methane trichloride, methyl trichloride, methenyl trichloride, trichloroform, "formyl trichloride.")

		Reference
Molecular weight:	119.38	
Melting point:	-63.5°C	K,J
Boiling point:	61.7°C	K
Liquid density: (see Figure 3.2)	1.483 g/ml (20°C)	K,C
Vapor specific gravity:	4.36 g/l (air = 1.0)	H

Vapor pressure: (see Figure 3.1)

Pres. (mm Hg)	Temp. (°C)	
61.0	0	H
100	10	H,K
~155	20	H,E
197.4	25	C
246.0	30	J
366.4	40	H

Solubility in water:

ppm (mg/l)	g/100g water	Temp. (°C)	
10,510	1.062	0	H
8,870	0.895	10	H
~8,100	0.822	20	A,H,E
7,800	--	25	A
~7,600	0.776	30	A,H

Partition coefficient: ($*C_s$ = saturation concentration)

Temp. (°C)	$*C_s$ air/C_s water (w/v per w/v)		
	Calc'd	Found	
20	--	0.12	E
20	0.13	0.12	A
25	0.16	0.13	A

Evaporation rates: (half-life)

	50% Evaporation Times				
Reference	A	A	A	B	Others in A
Avg. time, min.	20.2	25.7	18.5	21±4	34.5

90% evaporation times (Ref. C):
62 min., 68 min., 83 min. Avg. = 71

Conversion factors: (25°C, 1 atm in air)

1 ppm (v/v) = 0.0049 mg/l
1 mg/l = 204.8 ppm (v/v)
1 ppm (v/v) = 4.88 mg/m^3
1 mg/m^3 = 0.205 ppm (v/v)

Carbon Tetrachloride (CCl_4)

General Description. A clear, colorless, nonflammable liquid with a characteristic ethereal odor. It possesses excellent solvent properties for fats, oils, plastics, rubber, and many organic chemicals. It is practically insoluble in water, soluble in ethyl alcohol and acetone, and miscible in all proportions with diethyl ether, benzene, and chloroform. (Other names: methane tetrachloride, tetrachloromethane, perchloromethane, benzinoform.)

		Reference
Molecular weight:	153.82	
Melting point:	-22.99°C	K,C
Boiling point:	76.54°C	K,C
Liquid density: (see Figure 3.2)	1.594 g/ml (20°C)	K,C
Vapor specific gravity:	5.32 g/l (air = 1.0)	L,K

Vapor pressure: (see Figure 3.1)

Pres. (mm Hg)	Temp. (°C)	Reference
32.9	0.0	A
55.65	10.0	H
90	20.0	A,E
114.5	25.0	A,K
140	30.0	A
210.9	40.0	H

Solubility in water:

ppm (mg/l)	Temp. (°C)	Reference
785-800	20	A
800	25	H
814	30	A

Partition coefficient: ($*C_s$ = saturation concentration)

Temp. (°C)	$*C_s$ air/C_s water (w/v per w/v)		Reference
	Calc'd	Found	
25	1.20	0.87	A
25	0.97	0.91	A
20	--	0.91	E

Evaporation rates: (half-life)

	Evaporation times		90% Evaporation
Reference	A	B	B
Avg. time, min.	28.8	29	97

Conversion factors: (25°C, 1 atm in air)

1 ppm (v/v) = 0.0063 mg/l
1 mg/l = 158.9 ppm (v/v)
1 ppm (v/v) = 6.3 mg/l
1 mg/m^3 = 0.159 ppm (v/v)

BROMOMETHANES

Methyl Bromide (CH_3Br)

General Description. Methyl bromide is a colorless gas at normal temperatures and pressure. It is odorless in dilute concentrations and has a sweetish, chloroform-like odor at high concentrations. In the liquid phase, it is clear, colorless, and extremely volatile. It is practically nonflammable, but can form explosive mixtures under certain conditions. It is miscible in all proportions with ethyl alcohol, diethyl ether, chloroform, carbon disulfide, and most organic solvents. It is only slightly soluble in water.

			Reference
Molecular weight:	94.94		
Melting point:	-93.6°C		K,C
Boiling point:	3.56°C		K,C,H
Liquid density:	1.6755 g/ml (20°C)		K
Vapor density:	3.974 g/l (20°C)		J

Vapor pressure:

Pres. (mm Hg)	Temp. (°C)	
739.2	2.85	C
760	3.6	K
1,420	20.0	H,E
1,520	23.3	K
1,632.7	25.0	C

Solubility in water:

g/100 g water	Temp. (°C)	
1.75	20	H

Conversion factors:
(25°C, 1 atm in air)

1 ppm (v/v) = 0.0039 mg/l
1 mg/l = 257.5 ppm (v/v)
1 ppm (v/v) = 3.88 mg/m^3
1 mg/m^3 = 0.258 ppm (v/v)

Methylene Bromide (CH_2Br_2)

General Description. Methylene bromide is a colorless, heavy liquid that is slightly soluble in water and is miscible in all proportions with ethyl alcohol, diethyl ether, and benzene. (Other names: dibromomethane, methylene dibromide.)

		Reference
Molecular weight:	173.85	
Melting point:	-52.55°C	K
Boiling point:	97.0°C	K,F,I
Liquid density:	2.497 g/l (20°C)	K,H

Vapor pressure:

Pres. (mm Hg)	Temp. (°C)	
10	-2.4	K
40	+23.3	K
100	+42.3	K
400	+79.0	K

Solubility in water:

g/100g Water	Temp. (°C)	
1.173	0	G
1.146	10	G
1.148	20	G
1.176	30	G

Conversion factors:
(25°C, 1 atm in air)

1 ppm (v/v)	=	0.0071 mg/l
1 mg/l	=	140.6 ppm (v/v)
1 ppm (v/v)	=	7.11 mg/m^3
1 mg/m^3	=	0.141 ppm (v/v)

Bromoform ($CHBr_3$)

General Description. Bromoform is a colorless liquid with a strong chloroform-like odor and an agreeable taste. It is not an important solvent and it has relatively low volatility. It is slightly soluble in water, soluble in benzene and chloroform, and miscible in all proportions with ethyl alcohol and diethyl ether. (Other names: tribromomethane, methenyl tribromide.)

		Reference
Molecular weight:	252.75	
Melting point:	8.3°C	K,I
Boiling point:	149.5°C	K,H
Liquid density:	2.89 g/ml (20°C)	K,H,I

Vapor pressure:

Pres. (mm Hg)	Temp. (°C)	
10	34.0	K
40	63.6	K
100	85.9	K

Solubility in water:

g/100g Water	Temp. (°C)	
0.301	15	G
0.319	30	G

Conversion factors:
(25°C, 1 atm in air)

1 ppm (v/v) = 0.0103 mg/l
1 mg/l = 96.74 ppm (v/v)
1 ppm (v/v) = 10.33 mg/m^3
1 mg/m^3 = 0.097 ppm (v/v)

Carbon Tetrabromide (CBr_4)

General Description. Carbon tetrabromide is a solid in the form of monoclinic crystals or tablets with a peculiar odor. It is insoluble in water; slightly soluble in liquid hydrofluoric acid; soluble in ethyl alcohol, diethyl ether, and chloroform; and very soluble in carbon disulfide. (Other names: tetrabromomethane, carbon bromide.)

		Reference
Molecular weight:	331.65	
Melting point:	(α) 90.1°C	K,H,I
	(β) 94.3°C	K
Boiling point:	189.5°C	K,H,I
Density:	2.961 ($^{100}_{4}$) and 3.42 ($^{20}_{4}$)	K,H
Vapor pressure:	40 mm Hg at 96.3°C	K
Solubility in water:	0.024 g/100 g water (30°C)	G
Conversion factors: (25°C, 1 atm in air)	1 ppm (v/v) = 0.0136 mg/l 1 mg/l = 73.72 ppm (v/v) 1 ppm (v/v) = 13.56 mg/m^3 1 mg/m^3 = 0.0737 ppm (v/v)	

IODOMETHANES

Methyl Iodide (CH_3I)

General Description. Methyl iodide is a clear, heavy, volatile liquid with a sweet, acrid odor. It is usually colorless, but turns brown upon exposure to light. It has a high refractive index and is used in microscopy. It is slightly soluble in water, soluble in acetone and benzene, and miscible in all proportions with ethyl alcohol and diethyl ether. (Other name: iodomethane.)

			Reference
Molecular weight:	141.94		
Melting point:	−66.45°C		K,C,I
Boiling point:	42.4°C		K,C,I
Liquid density:	2.2790 g/ml (20°C)		H,K,C,I

Vapor pressure:

Pres. (mm Hg)	Temp. (°C)	Reference
100	−7.0	K
405.9	25.0	C
760	42.4	K
844	45.6	C

Solubility in water:

g/100g Water	Temp. (°C)	Reference
1.565	0	G
1.446	10	G
1.419	20	G
1.429	30	G

Conversion factors:
(25°C, 1 atm in air)

1 ppm(v/v)	=	0.0058 mg/l
1 mg/l	=	172.2 ppm (v/v)
1 ppm(v/v)	=	5.80 mg/m^3
1 mg/m^3	=	0.172 ppm (v/v)

Methylene Iodide (CH_2I_2)

General Description. Methylene iodide is a straw-colored or brown heavy liquid. It appears as yellow needles or leaf in the crystalline form. It is only slightly soluble in water but it is soluble in ethyl alcohol, diethyl ether, benzene, and chloroform. (Other name: diiodomethane.)

		Reference
Molecular weight:	267.84	
Melting point:	6.1°C	K,I
Boiling point:	182.0°C	K
Liquid density:	3.325 g/ml (20°C)	K,H,I
Solubility in water:	g/100g Water Temp. (°C)	
	0.124 30	G

Conversion factors:
(25°C, 1 atm in air)

1 ppm (v/v) = 0.0110 mg/l
1 mg/l = 91.28 ppm (v/v)
1 ppm (v/v) = 10.95 mg/m³
1 mg/m³ = 0.091 ppm (v/v)

Iodoform (CHI_3)

General Description. Iodoform is a yellow powder with a characteristic pungent, aromatic odor. It appears as yellow hexagonal prisms or needles in the crystalline form. It is insoluble in water and benzene and is soluble in hot ethyl alcohol, diethyl ether, acetone, chloroform, carbon disulfide, and acetic acid. (Other name: triiodomethane.)

Molecular weight:	393.73	
Melting point:	123°C	K
Boiling point:	~218°C	K
Density:	4.008 ($^{20}_{4}$)	K,H,I
Solubility in water:	0.01 g/100g Water (25°C)	H

Conversion factors:
(25°C, 1 atm in air)

1 ppm (v/v) = 0.0161 mg/l
1 mg/l = 62.10 ppm (v/v)
1 ppm (v/v) = 16.10 mg/m³
1 mg/m³ = 0.062 ppm (v/v)

MIXED HALOGENATED METHANES

Bromodichloromethane ($BrCHCl_2$)

General Description. Bromodichloromethane is a colorless liquid that is insoluble in water and very soluble in ethyl alcohol, diethyl ether, acetone, benzene, and most organic solvents. It is miscible in all proportions with chloroform.

		Reference
Molecular weight:	163.83	
Melting point:	-57.1°C	K
Boiling point:	90.0°C	K
Liquid density:	1.980 g/ml (20°C)	K
Conversion factors: (25°C, 1 atm in air)	1 ppm (v/v) = 0.0067 mg/l 1 mg/l = 149.2 ppm (v/v) 1 ppm (v/v) = 6.700 mg/m^3 1 mg/m^3 = 0.149 ppm (v/v)	

Dibromochloromethane (Br_2CHCl)

General Description. Dibromochloromethane is a colorless liquid that is insoluble in water and soluble in ethyl alcohol, diethyl ether, acetone, benzene, and most organic solvents.

Molecular weight:	208.29	
Melting point:	-22±	D
Boiling point:	119-120°C	K
Density:	2.451 ($^{20}_4$)	K
Conversion factors: (25°C, 1 atm in air)	1 ppm (v/v) = 0.0085 mg/l 1 mg/l = 117.4 ppm (v/v) 1 ppm (v/v) = 8.519 mg/m^3 1 mg/m^3 = 0.117 ppm (v/v)	

Dichloroiodomethane (Cl_2CHI)

General Description. Dichloroiodomethane is a liquid that is insoluble in water; soluble in diethyl ether, ethyl alcohol, acetone, and benzene; and very soluble in chloroform.

Molecular weight:	210.83	
Boiling point:	132°C	K
Density:	2.392 ($^{20}_4$)	K
Conversion factors: (25°C, 1 atm in air)	1 ppm (v/v) = 0.0086 mg/l 1 mg/l = 116.0 ppm (v/v) 1 ppm (v/v) = 8.623 mg/m^3 1 mg/m^3 = 0.116 ppm (v/v)	

KEY TO LETTERED REFERENCES

A Dilling, W.L. (1977) Interphase transfer processes. II Evaporation rates of chloromethanes, ethanes, ethylenes, propanes, propylenes from dilute aqueous solutions, comparisons with theoretical predictions. Environmental Science and Technology 11(4):405-409.

B Dilling, W.L., N.B. Tefertiller, and G.J. Kallos (1975) Evaporation rates and reactivities of methylene chloride, chloroform, 1,1,1-trichloroethane, trichloroethylene, tetrachloroethylene, and other chlorinated compounds in dilute aqueous solutions. Environmental Science and Technology 9:833-838.

C Dreisbach, R.R.(1961) Physical Properties of Chemical Compounds. Vols. II and III, No. 29 Advances in Chemistry Series. Washington, D.C.: American Chemical Society.

D Lange, N.A., ed. (1961) Handbook of Chemistry, 10th ed. New York, N.Y.: McGraw-Hill Book Company, Inc.

E McConnell, G., D.M. Ferguson, and C.R. Pearson (1975) Chlorinated hydrocarbons and the environment. Endeavor 34(121): 13-18.

F Rodd, E.H., ed. (1951) Chemistry of Carbon Compounds Vol. 1, Part A. Houston, Tex.: Elsevier Press, Inc.

G Seidell, A., (1941) Solubility of Organic Compounds, 3rd ed., Vol. II. New York, N.Y.: D. Van Nostrand Co., Inc.

H Standen, A., ed. (1964) Kirk-Othmer Encyclopedia of Chemical Technology, 2nd ed. New York, N.Y.: Interscience Publishers.

I Von Oettingen, W.F. (1955) The Halogenated Hydrocarbons: Toxicity and Potential Dangers, Public Health Service Publication No. 414. Washington, D.C.: U.S. Government Printing Office.

J Von Oettingen, W.F. (1964) The Halogenated Hydrocarbons of Industrial and Toxicological Importance. Amsterdam: Elsevier Publishing Co.

K Weast, R.C., ed. (1974) Handbook of Chemistry and Physics, 55th ed. Cleveland, Ohio: CRC Press, Inc.

L Yaws, C.L. (1976) Methyl chloride, methylene chloride, chloroform and carbon tetrachloride. Chemical Engineering 83(14):81-89.

APPENDIX B   DIRECT HEALTH EFFECTS OF
                          NONFLUORINATED HALOMETHANES

CONTENTS

MONOHALOGENATED METHANE DERIVATIVES 265

    Methyl Chloride 265
    Methyl Bromide 266
    Methyl Iodide 268

DIHALOGENATED METHANE DERIVATIVES 270

    Methylene Chloride 270
    Methylene Bromide 272
    Methylene Iodide 273

TRIHALOGENTAED METHANE DERIVATIVES 273

    Chloroform 273
    Bromoform 275
    Iodoform 276

TETRAHALOGENATED METHANE DERIVATIVES 277

    Carbon Tetrachloride 277
    Carbon Tetrabromide 281

ABBREVIATIONS

LC_{50} Median lethal concentration.

LD_{50} Median lethal dose.

LD_{100} Dose that will kill 100 percent of experimental subjects.

LCL_0 Lowest published lethal concentration.

LDL_0 Lowest published lethal dose.

TCL_0 Lowest published toxic concentration.

TDL_0 Lowest published toxic dose.

MONOHALOGENATED METHANE DERIVATIVES

METHYL CHLORIDE

General. Methyl chloride is not regarded as highly toxic, but many reports of poisoning exist. The major problem accounting for most poisonings is that the compound is colorless and essentially odorless and produces no perceptible irritation to the eyes. Because the victim is often unaware of its presence, exposures become sufficiently prolonged to produce toxic effects even though concentrations may be low. The characteristic latent period between time of exposure and onset of symptoms also serves to prolong exposures.

Absorption, Metabolism, Excretion. Absorption occurs primarily through the lungs. Absorption through the skin is still open to question and no poisonings have been reported as a result of ingestion. Von Oettingen et al. (1949) showed that upon exposure to vapor, the concentration of methyl chloride in the blood increases but reaches only a moderate level which does not increase much with continued exposure because of the rapidity with which the compound disappears from the blood. These authors found that only 5 percent of the total amount injected is excreted in bile and urine. Some decomposition reaction must occur in the body after absorption. Morgan et al. (1967) postulate that the low excretion rate may be due to the enzyme catalyzed methylation of erythrocyte sulfhydral groups (NRC 1977). Lewis (1948) similarly suggested that methyl chloride interferes with enzyme systems by reacting with the sulfhydral group, as was demonstrated for methyl bromide and methyl iodide.

Flury (1964) theorized that methyl chloride is hydrolyzed to hydrogen chloride and methanol, which in turn is oxidized to formaldehyde and formic acid. Possibly some conversion to methyl alcohol occurs, but the evidence indicates that this is unlikely.

Acute Toxic Concentrations. Dosages needed to produce toxic effects in man have not been clearly delineated. Von Oettingen (1964) reports that acute systemic poisonings may occur from inhalation exposure to concentrations well above 500 ppm. Scharnweber et al. (1974) report that cases of methyl chloride intoxication primarily have involved acute exposures to levels well in excess of the currently accepted U.S. Occupational Standard (Time Weighted Average, TWA) of 100 ppm, and usually at levels well above 500 ppm. The consensus seems to be that an exposure to a concentration of 500 ppm or greater is necessary for acute systemic poisoning to occur. The amount that must be ingested to produce toxic effects in humans is unknown.

Observations in Man. Methyl chloride is a central nervous system depressant that has the least narcotic action of the chlorinated methane derivatives. Its toxic manifestations are primarily neurological. Commonly reported symptoms are similar to those described by Hansen et al. (1953).

Light cases of methyl chloride intoxication are marked by a characteristic latent period of one half to several hours from the end of exposure and the onset of symptoms. Recovery usually occurs within five to six hours after removal from the exposure. Repeated or unduly prolonged exposures may lead to more severe cases of poisoning, which may take days or weeks for recovery.

Severe, non-fatal systemic poisonings also are marked by the characteristic latent period. Damage to the liver and renal injury are common. Neurological symptoms are severe and often dominate the clinical picture. Some of these central nervous system effects may be irreversible and post-recovery symptoms such as headache, nervousness, and insomnia may persist.

265

Fatal cases of poisoning develop a similar symptomology after the characteristic latent period of several hours. The patient passes into a coma, and death usually ensues within hours or days, depending on the intensity of the exposure. The primary cause of death appears to be cerebral and pulmonary edema associated with circulatory failure. Pathological findings in fatal methyl chloride poisonings are in agreement with the clinical picture and reveal congestion, edema, and hemorrhages in various organs, especially the lungs and brain. Baird (1954) reported detecting methyl chloride in all organs analyzed after death.

Observations in Animals. These toxic doses are listed in the Registry of Toxic Effects of Chemical Substances (U.S. DHEW 1975):

Rat	Inhalation LCL_0	3,000 ppm/4H
Mouse	Inhalation LCL_0	3,146 ppm/7H
Guinea Pig	Inhalation LCL_0	20,000 ppm/2H

Patty (1958) reports the following effects after single exposure inhalation: most animals die in a short time at 150,000-300,000 ppm; dangerous effects in 30-60 minutes at 20,000-40,000 ppm; no serious effects for up to 60 minutes at 7,000 ppm; and no effect for up to eight hours at 500-1,000 ppm (NRC 1977).

Carcinogenesis. Unknown.

Mutagenesis. Tardiff et al. (1976) report that methyl chloride has been shown to be mutagenic in *S. typhimurium* TA 100 (base-pair substitutions with overlap to frameshift mutations).

Teratogenesis. Unknown.

METHYL BROMIDE

General. Toxicologists regard methyl bromide as a highly toxic substance. Hine (1969) reports that it has been responsible for more deaths among occupationally exposed persons than all of the more widely publicized and highly toxic organic phosphate groups. Like methyl chloride, it has a characteristic latent period between the time of exposure and the onset of symptoms, and because it is difficult to detect its presence there may be prolonged exposures. However, methyl bromide is much more dangerous than methyl chloride because of its greater toxicity.

Absorption, Metabolism, Excretion. Inhalation is the usual route for systemic poisoning and absorption occurs primarily through the lungs. If methyl bromide is ingested, absorption can occur through the gastrointestinal tract, but this is rare. Absorption through the skin is still in question. Miller and Haggard (1943) showed that when methyl bromide is absorbed it is rapidly broken down, as evidenced by an increase in the nonvolatile bromide level in blood and tissues. Storage seems to occur only as bromides, and tissues rich in lipoid material are the main storage areas.

The normal background level of nonvolatile bromide in the blood is about 0 to 1.5 mg/100 ml as reported by Clarke et al. (1945) and Bennatt and Courtney (1948). Severe methyl bromide poisonings have been reported with nonvolatile bromine levels in the blood of 2.4 to 25 mg/100 ml. Fatal poisonings have been reported with blood bromide levels as low as 8.3 and 9.2 mg/100 ml and as high as 211.6 mg/100 ml. Collins (1965) noted that blood bromide levels usually are raised to the range of 1.0 to 40 mg/100 ml, which is much lower than the increase resulting from inorganic bromide poisoning (100-200 mg/100 ml).

Excretion of methyl bromide occurs partly through the lungs. Miller and Haggard (1943) determined after the intraperitoneal injection of 60 mg/kg body weight of methyl bromide into rats that 90 percent of the total injection was exhaled undecomposed in the first 30 minutes. Thus, excretion through the lungs is rapid. Clarke et al. (1945) showed that methyl bromide also is partially excreted as bromide in the urine. Although the initial excretion of methyl bromide is rapid, it may take considerable time before it is completely eliminated. This may account for the prolonged nature of some symptoms.

The mechanism of toxic action and the precise decomposition reaction within the body after absorption are unknown. Flury (1964) postulated that methyl bromide is hydrolyzed to form methanol and hydrogen bromide, but this seems unlikely. Miller and Haggard (1943) showed that methyl bromide readily passes through cell membranes and bromide ions do not. This suggests the occurrence of intracellular decomposition, which could support the theory by Lewis (1948) that methyl bromide, a strong methylating agent, undergoes a methylation reaction with the sulfhydryl group found in many enzymes which liberates hydrogen bromide that in turn can form bromides.

Rathus and Landy (1961) theorize that methyl bromide is first absorbed in extracellular fluids such as the blood, then it slowly diffuses into surrounding cells and reacts with the sulfhydrylic groups, interfering with cell metabolism by affecting the enzyme systems that control intracellular oxidative reactions (NRC 1977). The toxic action thus would depend on the concentration of conjugated methyl bromide complexes within the cells. In low concentrations the toxic effect would be irritative, but in higher concentrations the effect may be paralytic and irreversible.

Acute Toxic Concentrations. The doses required to produce toxic effects in humans can be estimated fairly reliably from the many case reports and other studies that have been done. Kubota (1955) reported that human fatalities can occur from air exposure to greater than 600 ppm methyl bromide and that levels of 100-150 ppm could be harmful. Johnstone (1945) reported 34 cases of systemic poisoning from exposure to 100-500 ppm methyl bromide. Rathus and Landy (1961) reported seven cases of acute systemic poisoning from exposure to air concentrations of 200-500 ppm. Fatal poisonings have been reported as a result of exposure to 300-400 ppm, 8,000 ppm, and 60,000 ppm by Bruhin (1943), Miller (1943), Wyers (1945), Tourangeau and Plamondon (1945), and Viner (1945). Collins (1965) and Clarke et al. (1945) report that inhalation of 10,000 ppm methyl bromide for a few minutes is fatal. Watrous (1942) reported 31 systemic poisonings as a result of two weeks exposure for 8 hours a day at about 35 ppm level. The Registry of Toxic Effects of Chemical Substances (U.S. DHEW 1975) reports a TCLo of 35 ppm in humans for methyl bromide and a TDLo of 8,000 ppm by skin contact. Collins (1965) noted neurological and psychiatric manifestations occurring at 35 ppm and Clarke et al. (1945) reported mild symptoms from eight hour exposure to 50 ppm methyl bromide. It appears that exposures to low concentrations in the range of 50-500 ppm methyl bromide can result in acute systemic poisoning and that symptoms increase in severity with increasing levels of exposure. Fatal poisonings generally occur as a result of short duration exposures to concentrations much greater than 500 ppm, but prolonged exposure to lower levels is sufficient to result in fatality. The amount of methyl bromide that must be ingested to produce toxic effects in humans is unknown.

Observations in Man. Methyl bromide has been reported to produce local toxic effects when in contact with the skin. Burns are characterized by excessive vesication and damage is caused primarily to peripheral nerve endings.

The numerous case reports of poisoning offer a good clinical picture. Most report symptoms similar to those described by Araki et al. (1971). The severity and exact symptoms vary with the intensity of exposure and the susceptibility of the individual. Non-fatal methyl bromide poisonings are usually the result of exposure to low concentrations, but such exposures are more likely to be prolonged and can result in severe poisoning. There is a latent period of from 2 to 48 hours, but usually there is about four to six hours between the exposure and the onset of symptoms.

Light cases of poisoning are marked by mild neurological symptoms that may be associated with gastrointestinal disturbances and seldom extend beyond the initial symptoms. Recovery takes only a few days.

Moderate cases of poisoning have the same initial symptoms but more neurological symptoms and visual disturbances develop, indicating further involvement of the central nervous system. Complete recovery may take weeks or months during which many symptoms may persist.

Severe cases of poisoning develop with the same characteristic latent period and initial symptoms. The most prominent physical signs are slow, hesitant, confused speech; staggering of gait; tremors; and uncoordination. Psychic disturbances are common. Tremors frequently pass into twitchings and finally to epileptiform convulsions. Recovery may take months and many symptoms may persist.

Fatal cases usually occur from single exposures to high concentrations. The patient usually experiences initial symptoms similar to those of light poisoning cases after a characteristic latent period of several hours, then tremors, and finally repetitive Jacksonian-type convulsions interspersed with periods of unconsciousness. The convulsions increase in intensity and frequency. Death occurs as a result of pulmonary edema within four to six hours after onset or occasionally as a result of circulatory failure after 24 to 48 hours. If the patient survives the first two days after onset, the prognosis is good.

The pathological changes in fatal methyl bromide poisoning are most characteristic and significant in the respiratory tract. The trachea and pleural cavity may contain quantities of blood-stained fluid. The lungs are hyperemic and edematous and may show consolidation. After the lungs, the brain shows the most characteristic changes. The meninges and the brain are hyperemic, edematous, and inflamed. Fatty degeneration in the ganglion cells of the brain, degeneration and necrosis of the renal tubular epithelium, central lobular necrosis of the liver, and extensive degeneration of the greater curvature of the stomach all have been reported. In fatal poisonings, the organ functions, especially the respiratory organs, are affected much more than in non-fatal poisonings and neurological symptoms are much less severe. Irritation of the central nervous system is not characteristic of most fatal exposures from high concentrations, indicating that the neurological changes are primarily of functional character. The immediate cause of death evidently is pulmonary edema associated with circulatory failure.

Observations in Animals. These toxic doses in animals are listed in the Registry of Toxic Effects of Chemical Substances (U.S. DHEW 1975):

Rat	Inhalation	LCL_0	514 ppm/6H
Rabbit	Inhalation	LCL_0	6,425 ppm/1H
Guinea Pig	Inhalation	LCL_0	300 ppm/9H

Carcinogenesis. Unknown.

Mutagenesis. Tardiff et al. (1976) report that methyl bromide has been shown to be mutagenic in $S.$ $typhimurium$ TA 100 (base-pair substitutions with overlap to frameshift mutations).

Teratogenesis. Unknown.

Conclusions. Methyl bromide is clearly a highly toxic agent that is widely used in the United States as an agricultural fumigant (Hine 1969). However, little firm evidence is available on its carcinogenic, mutagenic, or teratogenic potential.

METHYL IODIDE

General. Methyl iodide is regarded as a highly toxic substance by toxicologists but very little information is known about its toxicity because reports of poisoning in the literature are rare. It has strong narcotic and anesthetic properties.

Absorption, Metabolism, Excretion. Methyl iodide is absorbed through the lungs. If ingested, absorption can occur by way of the gastrointestinal tract. Absorption through the intact skin is still in question. Morgan et al. (1967) found that the retention of inhaled methyl iodide vapor is about 70 percent.

Little information exists on the fate of methyl iodide after absorption. Morgan et al. found that it is rapidly removed from the lungs and broken down, releasing iodide ions. Elevation of the blood serum iodide level and the cerebrospinal fluid iodide level have been reported. Morgan et al. also found that between 30-40 percent of the iodine introduced by methyl iodide will be accumulated by the thyroid. Methyl iodide is partially excreted as iodide in the urine. Complete excretion takes considerable time.

The mechanism of methyl iodide's toxic action is unknown. Lewis (1948) demonstrated that it has the same ability as methyl bromide to inhibit the action of certain enzyme systems within the cell by reacting with sulfhydrylic groups.

Acute Toxic Concentrations. The minimum toxic and fatal concentrations are not known. The only cases of poisoning reported are from exposures to unknown concentrations. Gosselin et al. (1976) estimate a probable lethal dose of 50 to 500 mg/kg. The U.S. Occupational Standard is set at a TWA maximum allowable concentration (MAC) of 5 ppm. Animal experiments indicate methyl iodide is the most toxic monohalogenated methane derivative.

Observations in Man. Methyl iodide has been reported to cause an acute toxic reponse as a result of skin contact. It is a vesicant, like methyl bromide, but it does not produce as severe an acute local response.

A number of acute methyl iodide poisonings have been reported and are reviewed by Appel et al. (1975) and Garland and Camps (1945). Poisoning symptoms develop after a characteristic latent period of several hours from the end of exposure. Cerebellar and neurologic symptoms are the most evident signs of impaired central nervous system function. If the result is not fatal, methyl iodide poisoning may take a long course and recovery may be incomplete, lasting up to four to six months. Psychiatric disturbances are prominent during the recovery period and may persist indefinitely.

Pathological findings indicate congestion of all organs, bronchopneumonia, and hemorrhages in the lungs. Death appears to be the result of massive pulmonary edema.

Observations in Animals. These are the toxic doses reported in the Registry of Toxic Effects of Chemical Substances (U.S. DHEW 1975):

Rat	Oral	LD_{50}	220 mg/kg
Rat	Subcutaneous	LD_{50}	110 mg/kg
Rat	Subcutaneous	TDL_0	50 mg/kg
Rat	Skin	LDL_0	800 mg/kg
Mouse	Inhalation	LCL_0	426 mg/kg
Mouse	Subcutaneous	LD_{50}	110 mg/kg

Carcinogenesis. Druckrey et al. (1970) reported local sarcomata in BD-strain rats after subcutaneous injections once weekly. Poirier et al. (1975) obtained increased incidences of lung tumors in A/Heston mice after injection i.p. 3 times weekly for a total of 24 doses at the MTD dose. Gribble (1974) reports that a single (50 mg/kg) or a weekly (10 mg/kg) subcutaneous injection produces massive local sarcomas in rats. However, tumors were not produced by oral or intravenous administration although the details of this were not reported.

Mutagenesis. Unknown.

Teratogenesis. Unknown.

Conclusions. The toxicity of the monohalomethanes is thought to increase from the chloride to the bromide to the iodide derivative, but the basic symptom complex is the same in all acute monohalomethane poisonings. Although there are insufficient data to estimate low dose effects, it is clear that methyl iodide is carcinogenic.

DIHALOGENATED METHANE DERIVATIVES

METHYLENE CHLORIDE

General. Methylene chloride is not regarded as especially toxic but poisonings have been reported. It is one of the least toxic chlorinated hydrocarbons and the degree of poisoning usually is limited to transient effects, especially narcosis.

Absorption, Metabolism, Excretion. Absorption occurs mainly through the lungs, but it may also occur through the gastrointestinal tract if methylene chloride is ingested. Absorption through the intact skin does occur to some extent.

Lehmann and Schmidt-Kehl (1936) found that about 70-75 percent of the inhaled vapors were absorbed by the human body. Riley et al. (1966) reported the absorption of methylene chloride as ranging from 70 percent at the beginning of exposure to 31 percent at the end of exposure in a subject exposed to 100 ppm of methylene chloride for two hours. DiVincenzo et al. (1972) reported absorption of 50-60 percent as a result of exposures to 100 ppm for four hours and 200 ppm for two hours. Astrand et al. (1975) found that subjects at rest absorbed 55 percent of the inhaled methylene chloride after 30-minute exposures to concentrations of 250 ppm and 500 ppm. The percent absorption decreased in subjects exposed during exercise, but because of the greater amount inhaled, the total amount of absorbed methylene chloride increased. In general, about 50-75 percent of the inhaled methylene chloride is retained in the organism and exertion during the exposure results in greater amounts being absorbed and a greater risk of severe poisoning.

Von Oettingen et al. (1949, 1950) showed that the methylene chloride concentration in the blood increases rapidly during inhalation until an equilibrium is achieved. The final level depends primarily upon the concentration of methylene chloride in the air. Von Oettingen et al. report a fairly uniform distribution of methylene chloride in the heart, liver, and brain after absorption.

Excretion of methylene chloride takes place mainly through the lungs. DiVincenzo et al. (1972) have shown that the amount of methylene chloride excreted in breath and blood is directly proportional to the exposure dose. They report that about 40 percent of the methylene chloride absorbed during the exposure is not eliminated through the lungs. Some methylene chloride is excreted undecomposed in the urine but these are small amounts and they cannot account for some 40 percent of the amount absorbed that is not excreted through the lungs. Some sort of decomposition reaction must occur within the body.

Experiments by Stewart et al. (1972a, 1972b) suggest that carbon monoxide may be a metabolite of methylene chloride because of the elevated carboxyhemoglobin (COHb) levels observed. COHb formation has also been demonstrated in rats, rabbits, and guinea pigs. COHb is formed from the interaction of carbon monoxide and hemoglobin. The affinity of carbon monoxide is more than 200 times greater than that of oxygen. Therefore, COHb is formed preferentially and the oxygen content of the blood decreases. This oxygen deprivation may cause serious permanent damage to the brain and heart. The results of animal experimentation by several investigators indicate that carbon dioxide, formaldehyde, and formic acid may be additional metabolites of methylene chloride.

Acute Toxic Concentrations. The minimum toxic and fatal concentrations of methylene chloride for humans have not clearly been established. NIOSH has proposed reducing the U.S. Occupational standard for air concentrations to a TWA of 75 ppm. The Registry of Toxic Effects of Chemical Substances (U.S. DHEW 1975) reports an inhalation TCL₀ of 500 ppm for humans for an eight hour period. Stewart et al. (1972a, 1972b) report that no overt illness symptoms develop at exposures of 213 ppm for one hour or at 514 ppm for one hour. After one hour exposure to 986 ppm of methylene chloride, two of the three subjects developed lightheadedness which cleared within five minutes after the exposure. Winneke (1974) reported that subjects exposed at concentrations of 317, 470, and 751 ppm for three to four hours showed decreased performance in most psychomotor tasks. Experiments indicate that central nervous system function is disturbed at levels in the range of 317 ppm to 986 ppm.

In general, neither central nervous system disturbances nor undesirable carboxyhemoglobin saturation levels are likely to occur with exposures up to 300 ppm for up to one hour. At exposures up to 500 ppm for less than one hour the central nervous system will not yet be affected but high carboxyhemoglobin levels may be observed. These CCHb levels are higher than the levels resulting from exposures to carbon monoxide at its threshold limit value. (Consequently, NIOSH has recommended that methylene chloride's threshold limit value be lowered.) The central nervous system is definitely disturbed at exposures to greater than 750 ppm for less than one hour. Symptoms of illness will develop at exposures of 2,500 ppm for less than one hour and at lower concentrations if the exposure is prolonged. The fatal concentration of methylene chloride appears to be slightly greater than 20,000 ppm. Dosages required to produce toxic effects by way of ingestion are unknown.

Observations in Man. Contact of methylene chloride vapors with the eyes, respiratory tract, and skin will cause local irritation.

The systemic effects of poisoning have been reported by Hughes (1954), Collier (1936), Moskowitz and Shapiro (1952), and Stewart and Hake (1976). Light cases of methylene chloride poisoning are characterized by somnolence, lassitude, anorexia, and mild lightheadedness. No other severe symptoms are noted and recovery is complete and rapid after the end of exposure. Severe cases of methylene chloride poisoning are characterized by more marked central nervous system depression and disturbance. No cases of permanent disability have been reported, indicating that recovery is complete. The pathological findings are limited and it is difficult to characterize methylene chloride poisoning. The primary cause of death is progressive heart failure due to cardiac injury.

Vozovaya et al. (1974) report that chronic inhalation of methylene chloride by pregnant women resulted in the passage of methylene chloride through the placental barrier into the fetal blood and tissues. Methylene chloride was also found in the milk of lactating workers five to seven hours after the start of a work shift.

Stewart and Hake (1976) and Scott (1976) present recent evidence that exposure to methylene chloride in paint and varnish removers has caused fatal heart attacks in several persons. At least one of these heart attack victims had no prior history of heart disease.

Observations in Animals. These toxic doses in animals are listed in the Registry of Toxic Effects of Chemical Substances (U.S. DHEW 1975):

Rat	Oral	LD_{50}	2,136 mg/kg
Mouse	Intraperitoneal	LD_{50}	1,500 mg/kg
Mouse	Subcutaneous	LD_{50}	6,460 mg/kg
Dog	Oral	LDL_o	3,000 mg/kg
Rabbit	Oral	LD_{Lo}	1,900 mg/kg
Dog	Intraperitoneal	LD_{Lo}	950 mg/kg
Dog	Subcutaneous	LDL_o	2,700 mg/kg
Dog	Intravenous	LDL_o	200 mg/kg
Rabbit	Subcutaneous	LDL_o	2,700 mg/kg
Guinea pig	Inhalation	LCL_o	5,000 ppm/2H

Carcinogenesis. Unknown.

Mutagenesis. Filippova et al. (1967) found methylene chloride not to be mutagenic in Drosophila (NRC 1977).

Teratogenesis. Schwetz et al. (1975) found extra sternebra in 50 percent of the litters from 19 rats and nine mice exposed for seven hours daily to 1,250 ppm methylene chloride on days 6 to 15 of gestation versus 14 percent of the litters from controls. Cleft pallets and rotated kidneys also were found in two of the litters from exposed females. There was an increased incidence of dilated renal pelvis and delayed ossification of the sternebra in the fetal rats. These findings indicate feto-toxicity but not teratogenicity.

Chronic Effects. Heppel et al. (1944) studied the effects of daily inhalation of methylene chloride on several species of animals. At 10,000 ppm for four hours/day, five days/week over eight weeks, two of four dogs exhibited moderate centrilobular congestion with narrowing of liver cell cords and slight to moderate fatty degeneration. Slight to moderate fatty degeneration of the liver was found in 4 of 6 guinea pigs at similar exposures.

Weinstein and Diamond (1972) examined the effects of 5,000 ppm continuous exposure for seven days on female mice. Within the first 24 hours, physical activity and food and water intake decreased and the animals became lethargic. After 96 hours normal activity resumed. Liver weights increased initially but returned to control values at the end of seven days but body weights decreased. On the seventh day, centrilobular fatty infiltration was found in the livers of exposed mice.

MacEwan et al. (1972) continuously exposed animals at either 1,000 or 5,000 ppm for 14 weeks. At 5,000 ppm, animals of all species became relatively inactive in the first two days. Many of the animals died from the exposure, which was discontinued after 35 days. Dogs that died during exposure showed fatty livers, icterus, pneumonia, splenic atrophy, and edema of the brain and meninges.

Haun et al. (1972) continuously exposed mice to 25 and 100 ppm for up to two weeks. The results indicate no significant differences in organ weights between exposed and control mice and changes in the liver were noted.

METHYLENE BROMIDE

General. Very little information is available concerning the toxicity of methylene bromide. Evidently it has not given rise to human poisoning because no reports of poisoning have been found.

Absorption, Metabolism, Excretion. Absorption most likely occurs primarily through the lungs but very little information is known. The toxic mechanism and the fate of methylene bromide after absorption are unknown.

Acute Toxic Concentrations. Minimal toxic and fatal concentrations of methylene bromide for humans are unknown. On the basis of animal experiments, the toxicity lies intermediate to methylene chloride and methylene iodide. Therefore, toxic concentrations probably also lie intermediate to those of these compounds.

Observations in Man. No reports of human methylene bromide poisoning.

Observations in Animals. This toxic dose for animals is reported in the Registry of Toxic Effects of Chemical Substances (U.S. DHEW 1975):

Mouse   Subcutaneous   LD$_{50}$ 3,738 mg/kg

Similar LD$_{50}$ values in mice indicate that methylene bromide is less toxic than bromoform, which in turn is less toxic than carbon tetrabromide.

Kutob and Plaa (1962) found that no hepatic damage was observed at dosages significantly greater than the LD$_{50}$ value and they classified methylene bromide as quick-acting, like bromoform, whereas carbon tetrabromide was classified as having a delayed action.

Dykan (1964) reports that rabbits exhibit central nervous system disturbances from acute exposures to 17-20 mg/l. Lucas (1928) noted marked signs of central nervous system disturbance such rabbits' inability to stand after rectal administration of 10 cc methylene bromide.

Pathological findings indicate marked congestion of lungs, and the liver and other tissues contained substantial amounts of inorganic bromide. Death appears to be the result of pulmonary edema.

Carcinogenesis. Unknown.

Mutagenesis. Unknown.

Teratogenesis. Unknown.

METHYLENE IODIDE

General. Available information concerning the toxicity of methylene iodide is even less extensive than that of methylene bromide.

Absorption, Metabolism, Excretion. Unknown.

Acute Toxic Concentrations. Unknown.

Observations in Man. Methylene iodide has apparently not resulted in any human poisonings. Sax (1975) states that details of the toxicity of methylene iodide are unknown but it is probably an irritant to the respiratory tract in high concentrations as well as having narcotic properties that result in central nervous system depression.

Observations in Animals. The Registry of Toxic Effects of Chemical Substances (U.S. DHEW 1975) reports the subcutaneous LD_{50} of methylene iodide for mice as 830 mg/kg body weight, which would make it the least toxic of the iodinated methane compounds.

Carcinogenesis. Unknown.

Mutagenesis. Unknown.

Teratogenesis. Unknown.

TRIHALOGENATED METHANE DERIVATIVES

CHLOROFORM

General. Chloroform is not regarded as a highly toxic material by toxicologists. It is considered several times more potent than carbon tetrachloride as a depressant of the central nervous system when inhaled, but less toxic than carbon tetrachloride when ingested.

Absorption, Metabolism, Excretion. Absorption of chloroform occurs rapidly through the lungs and through the gastrointestinal tract if ingested. Absorption through intact skin has been demonstrated in animal and human experiments. Pulmonary absorption in humans decreases with duration of the exposure. The chloroform level in the blood rises rapidly and then more slowly until an equilibrium is established between its concentration in the inhaled air and that in the blood after 80-100 minutes of exposure. Lehmann and Hasegawa (1910) found percent absorptions in the range of 54-73 percent from exposures of 2,700-6,500 ppm for 3 to 10 minutes. Exposure concentrations of 4,300-5,000 ppm of chloroform for 20 minutes resulted in 49.4-77 percent absorption.

After chloroform is absorbed, it is rapidly distributed to all organs of the body with the concentration varying depending on exposure level and duration. Chloroform is primarily excreted undecomposed through the lungs. This pulmonary excretion is slower than the absorption process and chloroform may persist for several hours or even days after exposure ends. Barrett et al. (1939) further showed that a small fraction of the absorbed chloroform also is excreted undecomposed by way of the urine. The total amount of chloroform absorbed cannot be accounted for by excretion processes, indicating that some degree of decomposition must occur within the body after absorption.

Precise decomposition reactions within the body and metabolic products have not been clearly established. Paul and Rubinstein (1963), by using carbon 14 labeled chloroform injected into rats, were able to show that 74 percent of the radioactivity appears as chloroform in the exhaled breath and that some appears in the form of carbon dioxide. Feynolds and Yee (1967) hypothesized that the hepatotoxicity of the chloroform is related to the binding of reduction products to the endoplasmic reticulum and to the formation of chloromethylated lipids and proteins in the liver.

Acute Toxic Concentrations. The dosages of chloroform required to produce certain toxic effects are fairly well established from the results of experimental exposures and case reviews of poisonings. In general, inhalation of chloroform vapors will cause moderate toxic effects at concentrations of about 1,000 ppm for a few minutes but higher concentrations rapidly become more toxic and prolonged exposures to 15,000 ppm may be dangerous to life. The U.S. Occupational Standard for air concentrations is proposed at a TWA of 50 ppm because chronic exposure to very low levels has resulted in systemic poisoning.

Von Oettingen (1964) reports that oral doses of 30 ml and 100 ml have resulted in severe, non-fatal poisonings and 200 ml has been fatal in human adults. Gosselin et al. (1976) report that the probable mean oral lethal dose is about one fluid ounce for a 150-pound person. In general, doses greater than 30 ml, or about one ounce, can cause serious or fatal poisonings.

Observations in Man. The reports of acute poisoning most often found in the literature are the result of accidental overdose during anesthesia or those of delayed poisoning following anesthesia. Chloroform also produces a local toxic response when in contact with intact skin because of its strong vesicant action. The symptomology of chloroform poisoning varies with respect to the route of administration and the severity of the exposure.

Cases of poisoning from the ingestion of chloroform are marked by gastrointestinal disturbances as well as narcosis. In severe cases, the victim usually passes into a coma immediately. This initial coma may last several hours and even though the victim may regain consciousness, experience delirium and irrationality, there is relapse into the coma and death. Extensive liver damage, with many contributing factors, is the cause of death.

Most chloroform poisonings result from inhalation of chloroform. It produces strong central nervous system depression and can result in complete narcosis. These poisonings are characterized by loss of sensation and abolition of motor functions. Death may result from respiratory failure because of paralysis of the respiratory center or from circulatory failure reducing the blood supply to the brain.

Delayed chloroform poisoning is characterized by a latent period of a few hours to a day before the patient experiences drowsiness, restlessness, jaundice, muscle tremors, cramps, nausea, and vomiting.

The pathological findings following chloroform poisoning are extensive. The outstanding pathological feature is the characteristic central lobular necrosis of the liver. The liver is the most notably affected organ but fatty degeneration of the kidneys and, to a lesser extent, the heart are also characteristic.

Observations in Animals. The Registry of Toxic Effects of Chemical Substances (U.S. DHEW 1975) lists the following toxic doses in animals:

Rat	Oral	LD_{50}	800 mg/kg
Rat	Inhalation	LC_0	8,000 ppm/4H
Mouse	Oral	LD_0	2,400 mg/kg
Mouse	Oral	TD_0	18 gm/kg/1200
Mouse	Inhalation	LC_{50}	28 ppm
Mouse	Subcutaneous	LD_{50}	704 mg/kg
Dog	Oral	LD_0	1,000 mg/kg
Dog	Inhalation	LC_{50}	100 ppm
Dog	Intravenous	LD_0	75 mg/kg
Rabbit	Inhalation	LC_{50}	59 ppm

		LD_{lo}	800 mg/kg
Rabbit	Subcutaneous		
Guinea pig	Inhalation	LC_{lo}	20,000 ppm/2H

Carcinogenesis. Eschenbrenner (1945) administered repeated oral doses of chloroform in olive oil to A-strain mice in a graded series of dosages that were non-necrotizing and necrotizing to the liver. Hepatomas were found in mice that received necrotizing doses one month after the last dose.

Page et al. (1976) found hepatocellular carcinoma and hepatic modular hyperplasia in male and female B6C3F1 strain mice given chloroform at 138 or 277 mg/kg and 238 or 477 mg/kg, respectively, five times a week for 78 weeks (IRC 1977). Kidney epithelial tumors were observed in male Osborne-Mendel rats given chloroform at 90 or 180 mg/kg five times a week for 78 weeks by the same investigators.

Mutagenesis. Unknown.

Teratogenesis. Schwetz et al. (1974b) exposed pregnant Sprague-Dawley rats to chloroform for seven hours daily on days 6 to 15 of gestation. At 30 ppm exposure, there was a significant incidence of wavy ribs and delayed skull ossification but no other effects when compared to controls. At 100 ppm exposure, there was a significant incidence of fetal abnormalities such as acuadia, imperforate anus, subcutaneous edema, missing ribs, and delayed skull ossification when compared to controls.

Thompson et al. (1974) administered oral doses of chloroform to rats and rabbits on days 6 to 18 of gestation. They found no evidence of teratogenicity at dosages up to 126 and 50 mg/kg, respectively, but the offspring of both species had reduced birth weights when compared to controls. This evidence indicates feto-toxicity, not teratogenicity.

Chronic Effects. Miklashevskii et al. (1966) exposed male guinea pigs and male albino rats to 0.4 mg/kg oral doses and 35 mg/kg ($1/50$ LD_{50}) and 125 mg/kg ($1/50$ LD_{50}) oral doses, respectively. The experiment ran for five months and daily administration was implied but not specified. Both species showed no changes in conditioned reflexes or in autonomic or cardiac activity, blood protein ratios, catalase concentrations, or phagocytic capacity at 0.4 mg/kg doses. The blood albumin-globulin ratio decreased and the blood catalase activity decreased by the second month of exposure in the guinea pigs given 35 mg/kg doses. Guinea pigs that died at this exposure showed fatty infiltration, necrosis, and cirrhosis of the liver parenchyma, as well as lipoid degeneration and proliferation of interstitial cells in the myocardium. The rats given 125 mg/kg doses showed impaired ability to develop new conditioned reflexes during the 4th and 5th month.

Conclusions. Chloroform is a strong central nervous system depressant which characteristically produces central lobular necrosis of the liver and to a lesser extent of the kidneys. Evidence indicates that it is carcinogenic in some species and also feto-toxic to a degree. The Registry of Toxic Effects of Chemical Substances (U.S. DHEW 1975) cites that systemic toxic effects are produced from chronic exposure to as low as 10 ppm chloroform.

BROMOFORM

General. Little is known about the toxicology of bromoform. There have been no reports of bromoform poisoning as a result of occupational exposure. Very little study has been conducted on its toxicology.

Generally, it is regarded as a highly toxic material. On the basis of LD_{50} values reported in the Registry of Toxic Effects of Chemical Substances (U.S. DHEW 1975), bromoform appears to be less toxic than both iodoform and chloroform. Bromoform also appears to be more toxic than methylene bromide but less toxic than carbon tetrabromide.

Absorption, Metabolism, Excretion. Absorption may take place through the lungs if the vapors are inhaled, from the gastrointestinal tract if it is ingested, and to a certain extent through the skin. Excretion of bromoform takes place partly through the lungs in the undecomposed state. Bromoform is at least partly decomposed within the body with the formation of bromide, which in turn is excreted in the urine. Complete excretion requires a relatively long period of time.

Acute Toxic Concentrations. The minimum toxic fatal concentrations of bromoform for humans have not been determined. The U.S. Occupational standard in air has been set at a time weighted average (TWA) of 0.5 ppm.

Observations in Man. Von Oettingen (1955) reports that exposure to bromoform vapors produces local irritation of the respiratory tract, irritation of the pharynx and the larynx, lacrimation and salivation. Most reported cases of bromoform poisoning are in foreign literature and all were the result of accidental overdoses administered to children for the treatment of whooping cough. Symptoms develop after a short latent period relative to that of other halomethane derivatives.

Light cases may exhibit listlessness, headache, and vertigo as the only toxic effects. In severe cases, unconsciousness is associated with loss of reflexes and occasionally convulsions. The primary cause of death is respiratory failure. Recovery from bromoform poisoning appears to occur rapidly, usually within several days, with no permanent disabilities.

The pathological findings indicate the presence of bromoform in all body organs. It appears that bromoform causes fatty degeneration and central lobular necrosis of the liver, which is characteristic of iodoform and chloroform poisoning.

Observations in Animals. The Registry of Toxic Effects of Chemical Substances (U.S. DHEW 1975) lists the following toxic doses in animals:

Mouse	Subcutaneous	LD_{50}	1,820 mg/kg
Rabbit	Subcutaneous	LD_0	410 mg/kg

Dykan (1962) found pathological changes in the kidney and liver of guinea pigs following the injection of 100-200 mg/kg of bromoform per day for 10 days (NRC 1977).

Carcinogenesis. Unknown.

Mutagenesis. Tardiff et al. (1976) report that bromoform has been shown to be mutagenic in $S.$ $typhimurium$ TA-100 (base-pair substitutions with overlap to frameshift mutations).

Stanford Research Institute (1976) found bromoform to be weakly mutagenic in Ames Salmonella (NRC 1977).

IODOFORM

General. Toxicologists do not regard iodoform as highly toxic but very few details of its toxicity are known. According to the reported LD_{50} values in the Registry of Toxic Effects of Chemical Substances (U.S. DHEW 1975), iodoform is more toxic than chloroform, which in turn is more toxic than bromoform.

Absorption, Metabolism, Excretion. Absorption of iodoform occurs through the lungs if the dust or powder is inhaled and through the gastrointestinal tract if it is ingested. Absorption also may occur through open wounds and after injection into tissues or joints, both of which are responsible for most of the poisonings that have been reported.

After iodoform is absorbed, it is broken down to a large extent by decomposition reactions with the liberation of iodine. The exact decomposition reactions have not been clearly identified. The distribution of iodoform within the body after absorption varies considerably with the type of administration and has not been clearly established.

Excretion of Iodoform is largely, if not entirely, through the urine in the form of iodide. A small portion may be excreted in the urine in the organic form. Mulzer (1905) demonstrated that the major part is excreted within 24 hours but complete excretion of the organic iodide may require three to four days and that of the inorganic iodide six to seven days. Only 60 percent of the iodine administered as iodoform could be accounted for in the urine.

Acute Toxic Concentrations. The minimum fatal concentration of iodoform in humans has not been established. The Registry of Toxic Effects of Chemical Substances (U.S. DHEW 1975) reports an oral TDL_0 of 114 mg/kg for humans. The toxic effect reported in this case was gastrointestinal tract irritation.

Observations in Man. Human iodoform poisoning occurs primarily as a result of some medicinal application of iodoform. The picture of iodoform poisoning varies with the route of absorption, the amount absorbed, and susceptibility of the individual.

Frauenthal (1891) reported on a patient who had accidently ingested eight grams of iodoform. The patient experienced gripping abdominal pain, severe diarrhea, and later a headache. Recovery was rapid and complete.

Bryan (1903) described an acute local dermatitis which develops after application of excessive iodoform. The dermatitis may last several weeks and is the result of existing or acquired hypersensitivity to iodoform.

Systemic poisoning can develop as a result of local application of iodoform to wounds. Central nervous system depression is common in light cases. Complete anorexia, complete apathy, and marked emotional instability characterize more severe cases. In extreme cases delirium, hallucinations, and delusions develop, culminating in coma. The symptoms of systemic iodoform poisoning usually develop after a characteristic latent period of 24 hours.

Pathological findings indicate fatty degeneration of the liver, kidney, and heart, and characteristic central lobular necrosis of the liver.

Observations in Animals. The Registry of Toxic Effects of Chemical Substances (U.S. DHEW 1975) lists the following toxic doses in animals:

Mouse	Subcutaneous	LD_{50}	629 mg/kg
Rabbit	Oral	LDI_0	450 mg/kg

Carcinogenesis. Unknown.

Mutagenesis. Unknown.

Teratogenesis. Unknown.

TETRAHALOGENATED METHANE DERIVATIVES

CARBON TETRACHLORIDE

General. Carbon tetrachloride is regarded as a highly toxic substance by toxicologists. It is one of the most harmful of the common solvents. Most fatal cases occur as a result of acute or subacute exposures.

Absorption, Metabolism, Excretion. Carbon tetrachloride is absorbed through the lungs and through the gastrointestinal tract. It is absorbed slower through intact human skin than several of the halogenated hydrocarbons, but its higher toxicity makes it the only one likely to absorb toxicologically significant amounts.

As with chloroform, the rate of absorption through the lungs decreases gradually with the duration of the exposure and finally reaches an equilibrium. The depressant effect of carbon tetrachloride on blood pressure, heart rate, and respiration rate is marked and prompt, so that the circulatory and respiratory changes it produces will interfere with its absorption and distribution. In general, Lehmann and Schmidt-Kehl (1936) found that 57-65 percent of the inhaled air concentration of carbon tetrachloride is absorbed following exposures at 650-1,600 ppm for 30 minutes.

The rate of absorption of carbon tetrachloride through the gastrointestinal tract is greatly affected by the diet. Patty (1963) reports that ethanol consumption and high-fat diets greatly enhance carbon tetrachloride uptake. In addition, drug use (especially barbiturates) and non-specific environmental stresses are known to potentiate carbon tetrachloride damage in mammals. Patty (1963) also reports that diets rich in certain vitamins or amino acids (sulfhydryl-containing compounds) may offer some protection from liver damage.

Little is known about the distribution of carbon tetrachloride within the human body after it is absorbed. As a rule, the organ distribution varies with the route of administration, its concentration, and the duration of the exposure. Carbon tetrachloride is primarily excreted through the lungs in the undecomposed state. Lehmann and Hasegawa (1910) found that 78.7 percent of the amount inhaled is excreted through the lungs during the subsequent six hours. Animal experiments indicate that when carbon tetrachloride is administered to mammals it is metabolized to some extent in the liver. This slow rate of degradation may be a primary factor in the toxicity of carbon tetrachloride from repeated low level exposures. The metabolites include chloroform, hexachloroethane, and carbon dioxide. However, most absorbed carbon tetrachloride is excreted promptly.

Paul and Rubinstein (1963) and Butler (1961) have postulated that the mechanism of metabolism and toxic action is the homolytic cleavage of the carbon-chloride bonds to yield free radicals, which can then alkylate the sulfhydral groups of many enzyme systems.

Acute Toxic Concentrations. The minimum concentrations of carbon tetrachloride that will produce harmful toxic effects in humans have been fairly well established as a result of human experimental exposures and reports of poisoning. The Registry of Toxic Effects of Chemical Substances (U.S. DHEW 1975) reports both an inhalation TCL$_0$ of 1,000 ppm and an inhalation TCL$_0$ of 20 ppm for humans. In general, levels of exposure below 300 ppm for less than 30 minutes will result in no overt symptoms of illness or signs of central nervous system depression. Prolonged exposure beyond 30 minutes will produce mild signs and symptoms of central nervous system depression at levels of exposure as low as 100 ppm. Levels of exposure below 100 ppm are tolerated up to 2.5 to 4 hours with no abnormal reactions, but prolonging the exposure, such as to an eight-hour workday or more, will result in the development of mild symptoms. Short exposures of only one to five minutes at concentrations ranging from 5,000-10,000 ppm will result in development of moderate symptoms. Loss of consciousness was reported to occur upon exposure to 14,000 ppm after only 50 seconds. Levels greater than 25,000 ppm of carbon tetrachloride in the air are considered to be dangerous to life.

The information on the toxic effects following the ingestion of different amounts of carbon tetrachloride is not as extensive. Von Oettingen (1964) reports that accidental ingestion of 14-20 ml has repeatedly caused fatal poisonings within three to five days; however, larger doses have occasionally not been fatal. Gosselin et al. (1976) report that the mean lethal dose of carbon tetrachloride for humans lies between 5-10 ml by oral administration, adding further that as little as 2 ml has caused death.

Observations in Man. Carbon tetrachloride has been shown to produce an acute local toxic effect when applied to intact human skin. It dissolves the protective fatty layer of the skin because of its solvent properties and may predispose the affected area to secondary infections.

A large number of poisonings have been reported as a result of accidental and suicidal ingestion of the liquid. Gosselin et al. (1976) report that there is a longer latent period after ingestion, about 24-36 hours, than the latent period of several hours after inhalation. Gastrointestinal and hepatorenal injuries are most prominent in poisoning from ingestion. Neurological symptoms usually develop later.

Most poisonings are the result of the inhalation of carbon tetrachloride vapors. Nervous symptoms predominate in inhalation poisoning, with initial symptoms including dizziness, giddiness, vertigo, and headache. Light cases

278

usually are restricted to the initial symptoms and recovery is prompt and complete after the exposure ends. Short duration, low-level exposures often produce symptoms of light carbon tetrachloride poisoning.

In all poisonings, the liver is the most drastically affected organ and signs of damage appear within a few days of onset. The injury to the liver normally begins first, but kidney injury may be more prominent clinically. Recovery from renal injuries may be prolonged. The circulatory system is affected relatively little in mild and moderate carbon tetrachloride poisoning. In cases of severe poisoning, the victim may have circulatory collapse. The primary cause of death in acute carbon tetrachloride poisoning is respiratory paralysis or circulatory failure.

Pathological changes in fatal carbon tetrachloride poisoning have been described in detail. The most prominent damage occurs in the liver and kidneys. The characteristic feature is centrilobular necrosis of the liver, which is also prominent in poisoning from chloroform, bromoform, and iodoform. The most prominent renal injury is the fatty degeneration and necrosis of the renal tubular epithelium of the nephrons. Other organs are affected but the changes are much less characteristic than those of the kidney and liver.

Observations in Animals. The Registry of Toxic Effects of Chemical Substances (U.S. DHEW 1975) lists the following toxic doses in animals:

Animal	Route	Measure	Value
Rat	Oral	LD_{50}	1,770 mg/kg
Rat	Inhalation	LCL_0	4,000 ppm/4H
Rat (neurological)	Subcutaneous	TDL_0	133 gm/kg/25 WI
Mouse (carcinogenic)	Oral	TDL_0	4,800 mg/kg/88 DI
Mouse	Inhalation	IC_{50}	9,526 ppm/8 H
Mouse	Intraperitoneal	LD_{50}	4,620 mg/kg
Mouse	Subcutaneous	LDL_0	3,200 mg/kg
Dog	Oral	LDL_0	1,000 mg/kg
Dog	Intravenous	LDL_0	125 mg/kg
Cat	Inhalation	LCL_0	38,110 ppm/2 H
Cat	Subcutaneous	LDL_0	4,785 mg/kg
Rabbit	Oral	LD_{50}	6,380 mg/kg
Rabbit	Inhalation	LCL_0	550 ppm
Rabbit	Subcutaneous	LDL_0	3,000 mg/kg
Guinea pig	Inhalation	LCL_0	20,000 ppm/2 H
Hamster (carcinogenic)	Oral	TDL_0	3,680 mg/kg/30 WI

Gardner et al. (1925) found centrilobular necrosis in dogs after ingestion, inhalation, subcutaneous and intraperitoneal injection, and rectal administration.

Carcinogenesis. Edwards (1941-42) found hepatomas in C3H and A strain mice fed 0.1 ml of a 40 percent solution of carbon tetrachloride in olive oil two to three times per week for a total of 23-58 feedings. Hepatomas were found in 126 of 143 C3H strain mice and in all of the 54 A strain mice.

Edwards et al. (1942-43) and Edwards and Dalton (1942-43) further reported finding hepatoma incidences of 60, 82, and 47 percent in Y, C, and L strain mice, respectively, after carbon tetrachloride treatment. Normal hepatoma incidences are less than 20 percent.

Eschenbrenner and Miller (1946) found hepatomas in all A strain mice given 120 daily liver necrotizing doses of carbon tetrachloride in olive oil administered by stomach tube. No hepatomas were found in mice receiving non-necrotizing doses.

Della Porta et al. (1961) found liver cell carcinomas in male and female Syrian golden hamsters following weekly doses of a 5 percent solution of carbon tetrachloride in corn oil for 30 weeks.

Reuber and Glover (1967) found hepatocellular carcinomas in Buffalo strain rats given 0.65 ml/kg doses, subcutaneously, twice a week for 12 weeks. The same investigators in 1970 reported finding liver cell carcinomas in Japanese, Osborne-Mendel, and Wistar rats administered doses of 1.3 ml/kg biweekly for 78, 105, and 68 weeks, respectively (Reuber and Glover 1970).

Rubin and Popper (1967) reported finding a number of cases of hepatomas developing in men several years after episodes of carbon tetrachloride poisoning.

IARC (1962) and Warwick (1971) have shown that carbon tetrachloride produces liver tumors in the mouse, hamster, and rat following several routes of administration, including inhalation and oral ingestion.

Mutagenesis. Barthelmes (1970) in his survey of mutagenic agents listed carbon tetrachloride as a chromosome-breaking agent. The National Science Foundation Panel on Manufactured Organic Chemicals in the Environment (1975) has reported some positive indications of the mutagenicity of carbon tetrachloride.

Teratogenesis. Smyth and Smyth (1935) and Smyth et al. (1936) observed the effects of chronic exposure to 50, 100, 200, or 1,000 ppm carbon tetrachloride in three generations of rats. No report of incidence of embryonic or fetal anomalies was made.

Schwetz et al. (1974a) studied the effects of 300 and 1,000 ppm seven hour daily exposures on pregnant Sprague-Dawley rats on days 6 to 15 of gestation. The fetuses showed no apparent anatomical abnormalities by gross examination. However, the incidence of microscopic sternebral anomalies was significantly increased at the higher exposure level. The investigators concluded that carbon tetrachloride is not teratogenic.

Chronic Effects. Prendergast et al. (1967) exposed 15 rats, 15 guinea pigs, 3 rabbits, 2 dogs, and 3 monkeys to 82 ppm of carbon tetrachloride for 8 hours/day, 5 days/week for a total of 30 exposures. The lungs of all species showed interstitial inflammation or pneumonitis and a high percentage of all species except dogs had a mottled appearance of the liver. Fatty changes were most severe in the livers of guinea pigs and early portal cirrhosis was apparent in this species.

The same investigators exposed similar animals to 10 ppm or 1 ppm for 90 days continuously. Weight losses or decreased growth occurred in all species and exposures (except for rats at 1 ppm). Exposure to 10 ppm resulted in a high incidence of enlarged and discolored livers in all species except dogs. Liver changes, including fibroplastic proliferation, collagen deposition, hepatic cell degeneration and regeneration, and alteration of the lobule structure were found in all species.

Adams et al. (1952) exposed guinea pigs and rats to 5, 10, 25, 100, 200, and 400 ppm for 7 hours/day, 5 days/week for up to 184 exposures over a period of 258 days. Guinea pigs showed increased liver weights at all concentrations, moderate amounts of fatty degeneration at 10 ppm or greater, and moderate liver cirrhosis at 25 ppm or greater. Rats showed increased liver weights at 10 ppm or greater and liver cirrhosis at 50 ppm or greater.

Adams et al. also exposed rabbits to 10, 25, 50 and 100 ppm. Decreased growth, increased kidney weights, and increased blood clotting times were observed at exposures to 50 and 100 ppm. Moderate fatty degeneration and liver cirrhosis developed after 178 exposures in 248 days at 25 ppm. Liver weights were increased at exposures to 25 ppm or greater.

Smyth et al. (1936) and Smyth and Smyth (1935) exposed monkeys, rats, and guinea pigs to various concentrations for 8 hours/day, 4-6 days/week for up to 321 days. Rats developed liver cirrhosis after 189 exposures at 50 ppm and evidence of interstitial cell proliferation and regeneration of liver cells was noted after 126 exposures at 50 ppm. Monkeys showed evidence of liver damage with marked increase in interstitial cells and cellular infiltration into septa after 62 exposures at 200 ppm.

Conclusions. The symptoms of acute carbon tetrachloride poisoning are characterized by gastrointestinal disturbances, neurological symptoms, and hepatorenal injury. Numerous carcinogenic bioassays have shown it to be carcinogenic in mice, rats, and hamsters. The mutagenic potential of carbon tetrachloride has been reported but not adequately demonstrated. Studies have indicated that carbon tetrachloride is not teratogenic but further investigation is necessary to adequately determine its teratogenic potential.

CARBON TETRABROMIDE

General.
Toxicologists for the most part are uncertain about the toxicity of carbon tetrabromide.

Absorption, Metabolism, Excretion.
Unknown.

Acute Toxic Concentrations.
The toxic concentrations for humans are not known. On the basis of animal toxicity studies, the threshold limit value (TLV) has been set at a TWA of 100 ppb in air, which recognizes its great toxicity with respect to other methane derivatives. Only methyl iodide has a lower subcutaneous LD_{50} value for mice.

Observations in Man.
No reported cases of poisoning.

Observations in Animals.
The Registry of Toxic Effects of Chemical Substances (U.S. DHEW 1975) lists the following toxic doses in animals:

Rat	Oral	LDL_0	1,000 mg/kg
Mouse	Subcutaneous	LD_{50}	298 mg/kg

This would make carbon tetrabromide more toxic than carbon tetrachloride and the most toxic of the brominated methane derivatives.

Rinz (1894) exposed several animals to carbon tetrabromide. Oral administration to a rabbit of 3 g suspended in water immediately resulted in central nervous system depression marked by respiratory arrest. Autopsy revealed severe irritation of the stomach and upper small intestines. Large amounts of carbon tetrabromide remained in the gastrointestinal tract and only traces of bromide were found in the urine. It appears that absorption through the gastrointestinal tract is slow and limited and that little decomposition occurs in the body.

Carcinogenesis.
Unknown.

Mutagenesis.
Unknown.

Teratogenesis.
Unknown.

REFERENCES

Adams, E.M., H.C. Spencer, V.K. Rowe, D.O. McCollister, and D.D. Irish (1952) Vapor toxicity of carbon tetrachloride determined by experiments on laboratory animals. Archives of Industrial Hygiene and Occupational Medicine 6:50-66.

Appel, G.B., R. Galen, J. O'Brien, and R. Schoenfeldt (1975) Methyl oxidide intoxification. A case report. Annals of Internal Medicine 82(4):534-536.

Araki, S., K. Ushio, K. Suwa, A. Akira, and K. Uehara (1971) Methyl bromide poisoning: a report based on fourteen cases. Japanese Journal of Industrial Health 13:507-513.

Astrand, I., P. Ovrum, and A. Carlsson (1975) Exposure to methylene chloride--I. Its concentration in alveolar air and blood during rest and exercise and its metabolism. Scandinavian Journal of Workplace and Environmental Health 1:78-94.

Baird, T.T. (1954) Methyl chloride poisoning. British Medical Journal (December 4) ii,1353.

Barrett, H.M., J.G. Cunningham, and J.H. Johnston (1939) Study of fate in organism of some chlorinated hydrocarbons. Journal of Industrial Hygiene and Toxicology 21:479-490.

Barthelmess, A. (1970) Mutagenic substances in the human environment. Pages 69-147, Chemical Mutagenesis of Mammals and Man, edited by G. Vogel and G. Rohrborn. Heidelberg, Germany: Springer-Verlag.

Benatt, A.J. and T.R.B. Courtney (1948) Uraemia in methyl bromide poisoning; case report. British Journal of Industrial Medicine 5:21-25.

Binz, C. (1894) Beitrage zur pharmakologischen kenntniss der halogene. Archiv fuer Experimentelle Pathologie und Pharmakologie 34:185-207. Cited in W.F. Von Oettingen 1955.

Bruhin, J. (1943) Deutsches Gesund Gerichtles Medicine 37, A253. Thesis, 1942. Zurich, Switzerland. (See Von Oettingen [1964] below.)

Bryan, W.A. (1903) Iodoform dermatitis. Journal of the American Medical Association 40:972-974.

Butler, T.C. (1961) Reduction of carbon tetrachloride in vivo and reduction of carbon tetrachloride and chloroform in vitro by tissues and tissue constituents. Journal of Pharmacology and Experimental Therapeutics 134:311-319.

Clarke, C.A., C.G. Roworth, and H.E. Holling (1945) Methyl bromide poisoning; account of four recent cases met with in one of H.M. ships. British Journal of Industrial Medicine 2:17-23.

Collier, H. (1936) Methylene dichloride intoxication in industry--A report of two cases. Lancet 1:594-595.

Collins, R.P. (1965) Methyl bromide poisoning: a bizarre neurological disorder. California Medicine 103:112-116.

Della Porta, G.D., B. Terracini, and P. Shubik (1961) Induction with carbon tetrachloride of liver cell carcinomas in hamsters. Journal of the National Cancer Institute 26:855-859.

DiVincenzo, G.D., F.J. Yanno, and B.D. Astill (1972) Human and canine exposures to methylene chloride vapor. American Industrial Hygiene Association Journal 33:125-135.

Druckrey, H., H. Kruse, R. Preussmann, S. Ivankovic, and C. Landschutz (1970) Cancerogene alkylierende substanzen. 3. Alkylhalogenide, -sulfate, -sulfonate und ringgespannte heterocyclen. Zeitschrift fur Krebsforschung und Klinische Onkologie 74:241-273.

Dykan, V.A. (1962) Changes in liver and kidney functions due to methylene bromide and bromoform. Nauchnye Trudy Ukrainskii. Nauchno-Issledovatel'skii Institut. Gigieny Truda i Profzabolevanii 29:82-90.

Dykan, V.A. (1964) Problems on toxicology, clinical practice, and work hygiene in the production of bromine-organic compounds. Gigiena (Kiev: Zdorove) Sb. pages 100-103.

Edwards, J.E. (1941-42) Hepatomas in mice induced with carbon tetrachloride. Journal of the National Cancer Institute 2:197-199.

Edwards, J.E. and A.J. Dalton (1942-43) Induction of Cirrhosis of the liver and hepatomas in mice with carbon tetrachloride. Journal of the National Cancer Institute 3:19-41.

Edwards, J.E., W.E. Heston, and A.J. Dalton (1942-43) Induction of the carbon tetrachloride hepatoma in strain mice. Journal of the National Cancer Institute 3:297-301.

Eschenbrenner, A.B. (1945) Induction of hepatomas in mice by repeated oral administration of chloroform, with obervations on sex differences. Journal of the National Cancer Institute 5:251-255.

Eschenbrenner, A.B. and E. Miller (1946) Liver necrosis and the induction of carbon tetrachloride hepatomas in strain A mice. Journal of the National Cancer Institute 6:325-341.

Filippova, L.M., O.A. Pan'shin, and R.G. Kostyanovskii (1967) Chemical Mutagens. IV. Mutagenic activity of geminal system. Genetika (8):134-148.

Flury, F. (1928) Archiv fuer Experimentelle Pathologie und Pharmakologie 138:75.

Frauenthal (1891) Two drachms of iodoform at a dose. New York Medical Journal 54:46.

Gardner, G.H., R.C. Grove, R.K. Gustafson, E.D. Maire, M.J. Thompson, H.S. Wells, and P.D. Lamson (1925) Studies on the pathological histology of experimental carbon tetrachloride poisoning. Bulletin of the Johns Hopkins Hospital 36:107-133.

Garland, A. and F.E. Camps (1945) Methyl iodide poisoning. British Journal of Industrial Medicine 2:209-211.

Gosselin, R.E., H.C. Hodge, R.P. Smith, and M.N. Gleason (1976) Clinical Toxicology of Commercial Products, 4th edition. Baltimore, Md.: Williams and Wilkins Co.

Gribble, G.W. (1974) Carcinogenic alkylating agents (letter to editor). Chemistry in Britain 10(1):101.

Hansen, H., N.K. Weaver, and F.S. Venable (1953) Methyl chloride intoxification; report of 15 cases. American Medical Association Archives of Industrial Hygiene and Occupational Medicine 8:328-334.

Haun, C.C., E.H. Vernot, K.I. Darmer, Jr., and S.S. Diamond (1972) Continuous animal exposure to low levels of dichloromethane. Pages 199-208, Proceedings of the 3rd Annual Conference on Environmental Toxicology. AMRL-TR-130, Paper No. 12. Wright-Patterson Air Force Base, Ohio: Aerospace Medical Research Laboratory.

Heppel, L.A., P.A. Neal, T.L. Perrin, M.L. Orr, and V.T. Porterfied (1944) Toxicology of dichloromethane--I. Studies on effects of daily inhalation. Journal of Industrial Hygiene and Toxicology 26:8-16.

Hine, C.H. (1969) Methyl bromide poisoning. A review of ten cases. Journal of Occupational Medicine 11:1-10.

Hughes, J.P. (1954) Hazardous exposure to some so-called safe solvents. Journal of the American Medical Association 156:234-237.

International Agency for Research on Cancer (1972) IARC Monograph on the Evaluation of Carcinogenic Risk of Chemicals to Man 1:53-60. Lyon: World Health Organization.

Johnstone, R. (1945) Methyl bromide intoxication of large group of workers. Industrial Medicine 14:495-497.

Kubota, S. (1955) Industrial poisoning in chemical plants. I. Methyl bromide poisoning. Journal of the Society of Organic and Synthetic Chemistry 13:605-606.

Kutob, S.D. and G.L. Plaa (1962) Assessment of liver function in mice with bromosulphalein. Journal of Applied Physiology 17:123-125.

Lehmann, K.B. and Hasegawa (1910) Studies of the absorption of chlorinated hydrocarbons in animals and humans. Archiv fuer Hygiene 72:327-342.

Lehmann, K.B. and L. Schmidt-Kehl (1936) The thirteen most important chlorinated aliphatic hydrocarbons from the standpoint of industrial hygiene. Archives of Hygiene 116:131-268. (U.S. Department of NEW Publication No. (NIOSH) 76-138 and 76-133.)

Lewis, S.E. (1948) Inhibition of SH enzymes by methyl bromide. Nature 161:692-693.

Lucas, G.H.W. (1928) A study of the fate and toxicity of bromine and chlorine containing anesthetics. Journal of Pharmacology and Experimental Therapeutics 34:237.

MacEwan, J.D., E.H. Vernot, and C.C. Haun (1972) Continuous animal exposure to dichloromethane. AMRL-TR-72-28, Systemed Corporation Report No. W-T1005. Wright-Patterson Air Force Base, Ohio: Aerospace Medical Research Laboratory.

Miklashevskii, V.E., V.N. Tugarinova, N.L. Rakhmanina, and G.P. Yakovleva (1966) Toxicity of chloroform administered perorally. Hygiene and Sanitation (USSR) 31:320-322. (U.S. Department of Health, Education, and Welfare Publication No. [NIOSH] 75-114.)

Miller, D.P. and H.W. Haggard (1943) Intracellular penetration of bromide as feature in toxicity of alkyl bromides. Journal of Industrial Hygiene and Toxicology 25:423-433.

Miller, J.W. (1943) Fatal methyl bromide poisoning. Archives of Pathology 36:505-507.

Morgan, A., D.J. Morgan, J.C. Evans, and B.A.J. Lister (1967) Studies on the retention and metabolism of inhaled methyl iodine. II. Metabolism of methyl iodide. Health Physics 13:1067-1074.

Moskowitz, S. and H. Shapiro (1952) Fatal exposure to methylene chloride vapor. American Medical Association Archives of Industrial Hygiene and Occupational Medicine 6:116-123.

Mulzer, P. (1905) Ueber des verhalten des iodoforms im thierkorper. Zeitschrift fur Experimentelle Pathologie und Therapie 1:446-479.

National Research Council (1977) Drinking Water and Health. Committee on Safe Drinking Water, Subcommittee on Organic Contaminants. Washington, D.C.: National Academy of Sciences.

National Science Foundation Panel on Manufactured Organic Chemicals in the Environment (1975) SRI data. In Johns (1976) Air Pollution Assessment of Carbon Tetrachloride. Mitre Technical Report MTR-7144. McLean, Va.: The Mitre Corp.

Page, N.P. and U. Saffiotti (1976) Report on Carcinogenesis Bioassay of Chloroform. March 1, 1976. Bethesda, Md.: Carcinogenesis Program, Division of Cancer Cause and Prevention, National Cancer Institute.

Patty, F.A., ed. (1963a) Industrial Hygiene and Toxicology, 2nd edition, Vol. II. New York: Interscience Publishers. (Refer to chapter by Clayton et al.)

Patty, F.A., ed. (1963b) Pages 1264-1268, Industrial Hygiene and Toxicology, 2nd edition, Vol. II. New York: Interscience Publishers.

Paul, B.B. and D. Rubinstein (1963) Metabolism of carbon tetrachloride and chloroform by the rat. Journal of Pharmacology and Experimental Therapeutics 141:141-148.

Poirier, L.A., G.D. Stoner, and M.B. Shimkin (1975) Bioassay of alkyl halides and nucleotide base analogs by pulmonary tumor response in strain A mice. Cancer Research 35:1411-1415.

Prendergast, J.A., R.A. Jones, L.J. Jenkins, and J. Siegel (1967) Effects on experimental animals of long term inhalation of trichloroethylene, carbon tetrachloride, 1,1,1-trichloroethane, dichlorodiflouromethane, and 1,1,-dichloroethylene. Toxicology and Applied Pharmacology 10:270-289.

Rathus, E.M. and P.J. Landy (1961) Methyl bromide poisoning. British Journal of Industrial Medicine 18:53-57.

Reuber, M.D. and E.L. Glober (1970) Cirrhosis and carcinoma of the liver in male rats given subcutaneous carbon tetrachloride. Journal of the National Cancer Institute 44:419-423.

Reuber, M.D. and E.L. Glover (1967) Hyperplastic and early neoplastic lesions of the liver in Buffalo strain rats of various ages given subcutaneous carbon tetrachloride. Journal of the National Cancer Institute 38:891-895.

Reynolds, E.S. and A.G. Yee (1967) Liver parenchymal cell injury--V. Relationships between patterns of chloromethane-C14 incorporation into constituents of liver in vivo and cellular injury. Laboratory Investigation 16:591-603.

Riley, E.C., D.W. Fassett, and W.L. Sutton (1966) Methylene chloride vapor in expired air of human subjects. American Industrial Hygiene Association Journal 27:341-348.

Rubin, E. and H. Popper (1967) The evaluation of human cirrhosis deduced from observations in experimental animals. Medicine 46:163.

Sax, N.I. (1975) Dangerous Properties of Industrial Materials, 4 ed. New York: Van Nostrand Reinhold Publishing Co.

Scharnweber, H.C., G.N. Spears, and S.R. Cowles (1974) Case reports. Chronic methyl chloride intoxication in six industrial workers. Journal of Occupational Medicine 16:112-113.

Schwetz, B.A., B.K.J. Leong, and P.J. Gehring (1974a) Embryo-and fetotoxicity of inhaled carbon tetrachloride, 1,4-dichloroethane and methyl ethyl Ketone in rats. Toxicology and Applied Pharmacology 28:452-464.

Schwetz, B.A., B.K.J. Leong, and P.J. Gehring (1974b) Embryo-and fetotoxicity of inhaled chloroform in rats. Toxicology and Applied Pharmacology 28:442-451.

Schwetz, B.A., B.K.J. Leong, and P.J. Gehring (1975) The effect of maternally inhaled trichloroethylene, perchloroethylene, methyl chloroform, and methylene chloride on embryonal and fetal development in mice and rats. Toxicology and Applied Pharmacology 32:84-96.

Scott, R. (1976) Household product safety. The Washington Post. p. Va. 1, September 9.

Smyth, H.F., H.F. Smyth, Jr., and C.P. Carpenter (1936) The chronic toxicity of carbon tetrachloride--Animal exposures and field studies. Journal Industrial Hygiene and Toxicology 18:277-298.

Smyth, H.F. and H.F. Smyth, Jr. (1935) Investigation of the Chronic Toxicity of Carbon Tetrachloride. Final Report to the producers' Committee. (U.S. Department of Health, Education, and Welfare Publication No. [NIOSH] 76-133.)

Stanford Research Institute Report (1976) Project No. LSC-4378, April 1, 1976. Menlo Park, Calif.: Stanford Research Institute.

Stewart, R.D., T.N. Fisher, M.J. Hosko, J.E. Peterson, E.D. Baretta, and H.C. Dodd (1972a) Experimental human exposure to methylene chloride. Archives of Environmental Health 25:342-348.

Stewart, R.D., T.N. Fisher, M.J. Hosko, J.E. Peterson, E.D. Baretta, and H.C. Dodd (1972b) Carboxyhemoglobin elevation after exposure to dichloromethane. Science 176:295-296.

Stewart, R.D. and C.L. Hake (1976) Paint-remover hazard. Journal of American Medical Association 235:398-401.

Tardiff, R.G., G.P. Carlson, and V. Simmon (1976) Halogenated organics in tap water: a toxicological evaluation. Pages 213-227, Proceedings of the Conference on the Environmental Impact of Water Chlorination, Robert L. Jolley, editor, Oak Ridge, Tennessee, October 22-24, 1975. NTIS CONF-751096. Springfield, Va.: National Technical Information Service.

Thompson, D.J., S.D. Warner, and V.B. Robinson (1974) Teratology studies on orally administered chloroform in the rat and rabbit. Toxicology and Applied Pharmacology 29:123. (Abstract)

Tourangeau, F.J. and S.R. Plamondon (1945) Cases of exposure to methyl bromide vapours. Canadian Journal of Public Health 36:362-367.

U.S. Department of Health, Education, and Welfare (1975) Registry of Toxic Effects of Chemical Substances. Washington, D.C.: U.S. Department of Health, Education, and Welfare.

Viner, N. (1945) Methyl bromide poisoning; new industrial hazard. Canadian Medical Association Journal 53:43-45.

Von Oettingen, W.F., C.C. Powell, N.E. Sharpless, W.C. Alford, and L.J. Pecora (1949) Relation between the toxic action of chlorinated methanes and their chemical and physiochemical properties. (National Institutes of Health Bulletin No. 191.)

Von Oettingen, W.F., C.C. Powell, N.E. Sharpless, W.C. Alford, and L.J. Pecora (1950) Comparative studies of the toxicity and pharmacodynamic action of chlorinated methanes with special reference to their physical and chemical characteristics. Archives Internationales de Pharmacodynamie et de Therapie 81(1):17-34.

Von Oettingen, W.F. (1955) The Halogenated Hydrocarbons: Toxicity and Potential Dangers. Public Health Service No. 414. Washington, D.C.: U.S. Government Printing Office.

Von Oettingen, W.F. (1964) The Halogenated Hydrocarbons of Industrial and Toxicological Importance. Amsterdam: Elsevier Publishing Co.

Vozovaya, M.A., L.K. Malyarova, and R.M. Enikeeva (1974) Methylene chloride content in maternal and fetal tissue during pregnancy and nursing of female workers at a rubberware plant. Gigiena Truda i Professional'nye Zabolevaniya 4:42-43.

Warwick, G.P. (1971) Metabolism of liver carcinogens and other factors influencing liver cancer induction. Pages 121-157, Liver Cancer. Albany, N.Y.: Q Corporation.

Watrous, R.M. (1942) Methyl bromide; local and mild systemic toxic effects. Industrial Medicine 11:575-579.

Weinstein, R.S. and S.S. Diamond (1972) Hepatotoxicity of dichloromethane with continuous inhalation exposure at a low dose level. Pages 209-222, Proceedings of the 3rd Annual Conference on Environmental Toxicology. AMRL-72-130, Paper No. 13. Wright-Patterson Air Force Base, Ohio: Aerospace Medical Research Laboratory.

Winneke, G. (1974) Behavioral effects of methylene chloride and carbon monoxide as assessed by sensory and psychomotor performance. Pages 130-144, Behavioral Toxicology-Early Detection of Occupational Hazards, edited by C. Xintaras, B.L. Johnson, and I. deGroot. (U.S. Department of Health, Education, and Welfare. Publication No. [NIOSH] 74-126.)

Wyers, H. (1945) Methyl bromide intoxication. British Journal of Industrial Medicine 2:24-29.

APPENDIX C   BIOGRAPHICAL SKETCHES
                                    OF PANEL MEMBERS AND
                                    PANEL ADVISOR

JULIAN B. ANDELMAN is a Professor of Water Chemistry in
the Graduate School of Public Health at the University of
Pittsburgh. He received an A.B. from Harvard College in
1952 and a Ph.D. in chemistry from the Polytechnic Institute
of Brooklyn in 1960. Dr. Andelman is an expert in and has
authored articles in the field of physical and water
chemistry specializing in the chemistry of trace materials
in natural waters and treatment systems, and their
significance for human health. He worked at the Bell
Telephone Laboratories in 1961-1963, was a visiting member
of the faculty at University College London, in 1970-71, and
has served as consultant to the World Helath Organization

JOHN H. CUMBERLAND is Director of the Bureau of Business
and Economic Research and Professor of Economics at the
University of Maryland. He received a B.A. from the
University of Maryland in 1947 and a M.A. and Ph.D. from
Harvard University in 1949 and 1951 respectively. He is a
specialist in environmental analysis, natural resource
economics, and regional and urban development. Dr.
Cumberland has authored two books and numerous articles in
these fields. He has served as consultant to various state,
federal, and international agencies including the
Environmental Protection Agency, Department of Commerce,
Department of Interior, National Aeronautics and Space
Administration, the World Health Organization, and
Organization of Economic Cooperation and Development, the
World Bank, and the International Institute of Applied
Systems Analysis.

EDWARD D. GOLDBERG is a Professor of Chemistry in the
Geological Research Division at Scripps Institution of
Oceanography, University of California at San Diego. He
received a B.S. from the University of California in 1942
and a Ph.D. in chemistry from the University of Chicago in
1949. Dr. Goldberg is an expert in the field of marine
geochemistry specializing in atmospheric and marine
pollution, low molecular weight halogenated hydrocarbons in
the ocean, and marine sedimentation. He served as a
Guggenheim fellow in Berne during 1961, and as a NATO fellow
at the University of Brussels in 1970, as well as being a
member of numerous advisory committees and author of
numerous articles.

DAVID G. HOEL is a mathematical statistician and Chief of the Environmental Biometrics Branch at the National Institute of Environmental Health Sciences. He received an A.B. from the University of California, Berkeley, in 1961 and a Ph.D. in statistics from North Carolina, Chapel Hill in 1966. He is an expert in the fields of biostatistics, statistical inference, and risk extrapolation with a specialty in carcinogens and organics in drinking water. Dr. Hoel has worked at Oak Ridge National Laboratory, served on advisory committees to government agencies, and authored numerous articles.

ROBERT J. MOOLENAAR is Director of Environmental Science Research at Dow Chemical Company. He received a B.A. from Hope College in 1953 and a Ph.D. from the University of Illinois in 1957. Dr. Moolenaar's expertise is in the fields of physical chemistry and environmental sciences with a specialty in inorganic chemistry, biodegradation, photodegradation, toxicology, and fates of pollutants. He taught physical chemistry at the University of Illinois in 1956 and 1957 and served as a research chemist at Dow from 1957 on.

REINHOLD A. RASMUSSEN was a Professor and Section Head in the Air Resources Section of the Chemical Engineering Branch at Washington State University when this study began; in September 1977 he moved to the Oregon Graduate Center. He received a B.S. from the University of Massachusetts in 1958, an M.Ed. and a Ph.D. in botany from Washington University in 1960 and 1964 respectively. Dr. Rasmussen's specialty is in atmospheric biochemistry and the role of naturally occurring organic volatiles in the atmosphere, especially chemical identification, photochemistry, and biological interactions with the environment and man. He has authored numerous articles and served as advisor and consultant to various government agencies and committees. Much of his responsibility in the Panel's work, including writing, attending meetings, and interacting with NAS/NRC staff was shared with a colleague with similar expertise, DAVID PIEROTTI.

DAVID PIEROTTI held a research position in the Air Resources Section of the Chemical Engineering Branch at Washington State University through June 1977 following which he moved to the Oregon Graduate Center. He received a B.A. in chemistry and mathematics from the University of California, Santa Cruz in 1973 and a M.S. in chemistry from the University of California, San Diego in 1974. His specialty is in atmospheric biochemistry and the role of naturally occurring organic volatiles in the atmosphere.

Mr. Pierotti has published numerous articles. He served as an advisor to the Panel and contributed equally with the other panel members toward completion of this study.

APPENDIX D   UNITS OF WEIGHT AND CONCENTRATION[a]

Description and Unit	Equivalent
Weight:	
1 metric ton (t)	1,000 kilograms
1 kilogram (kg)	1,000 grams
1 milligram (mg)	10^{-3} g
1 microgram (µg)	10^{-6} g
1 nanogram (ng)	10^{-9} g
1 picogram (pg)	10^{-12} g
Concentration:	
Weight: weight basis (for foods, body organs, and other solids)	
1 part per million (ppm)	1 mg/kg or 1 µg/g
1 part per billion (ppb)	1 µg/kg or 1 ng/g
1 part per trillion (ppt)	1 ng/kg or 1 pg/g
Weight: volume basis (for water	
1 mg/liter	approximately 1 µg/g on weight: weight basis
1 µg/liter	approximately 1 ng/g on weight: weight basis
1 ng/liter	approximately 1 pg/g on weight: weight basis
Weight: volume basis (for air)	
1 mg/cubic meter	10^3 µg/cubic meter or 10^6 ng/cubic meter

[a] The conversion of a weight: volume relationship to a volume: volume basis depends on the molecular weight of the dispersed substance. At 25°C and 760 mm Hg pressure, the conversion formula is: ppm by volume = mg/cubic meter X 24.45/molecular weight where mg/cubic meter is the measured concentration of the compound.